Dimensions of Human Geography

Dimensions of Human Geography

Debashish Banerjee

RANDOM PUBLICATIONS
NEW DELHI - 110 002 (INDIA)

Dimensions of Human Geography

ISBN 978-93-51112-67-9

Published in 2014 in India by

RANDOM PUBLICATIONS

4376-A/4B, Gali Murari Lal, Ansari Road
New Delhi-110 002
Phone: +9111-43580356, 23289044
E-mail: randomexports@gmail.com; sales@randompublications.com; info@randompublications.com

Reprinted 2024

Type Setting by: Friends Media, Delhi-110089
Digitally Printed at: Replika Press Pvt. Ltd.

Preface

Human geography is one of the two major branches of geography and is often called cultural geography. Human geography is the study of the many cultural aspects found throughout the world and how they relate to the spaces and places where they originate and then travel as people continually move across various areas. Some of the main cultural phenomena studied in human geography include language, religion, different economic and governmental structures, art, music, and other cultural aspects that explain how and/or why people function as they do in the areas in which they live. Globalization is also becoming increasingly important to the field of human geography as it is allowing these specific aspects of culture to easily travel across the globe. Cultural landscapes are also important because they link culture to the physical environments in which people live. This is vital because it can either limit or nurture the development of various aspects of culture. For instance, people living in a rural area are often more culturally tied to the natural environment around them than those living in a large metropolitan area. This is generally the focus of the "Man-Land Tradition" in the Four Traditions of geography and studies human impact on nature, the impact of nature on humans, and people's perception of the environment.

Geographical knowledge, both physical and social, has a long history. In the history of geography, geographers have often recorded and described features of the Earth that might now be considered the remit of human, rather than physical, geographers. For example Hecataeus of Miletus, a geographer and historian in ancient Greece, described inhabitants of the ancient world as well as physical features. Human geography developed out of the University of California, Berkeley and was led by Carl Sauer. He used landscapes as the defining unit of geographic study and said that cultures develop because of the landscape but also help to develop the landscape as well. In addition, his work and the cultural geography of today is highly qualitative rather than quantitative - a main tenant of physical geography. Today, human geography is still practiced and more specialized fields within it such as feminist geography, children's geography, tourism studies, urban geography, the geography of sexuality and space, and political geography have developed to further aid in the study of cultural practices and human activities as they relate spatially to the world.

This book explores the latest advances in the field of this subject. The subject matter, both as regards the arrangement of chapters as well as contents is designed to meet the requirement of the students in several Universities.

I thank all members of my team who have helped in the preparation of the book. My special thanks go to "Random Publications" who have published the book.

—Debashish Banerjee

Contents

1

Introduction

Human geography is one of the two major subfields of geography. Human geography is the study of human use and understanding of the Earth and the process which have affected this. It is linked to both social science and the humanities. Human geography broadly differs from physical geography in that it has a greater focus on studying intangible or abstract patterns surrounding human activity and is more receptive to qualitative research methodologies. It encompasses human, political, cultural, social and economic aspects of the social sciences. While the major focus of human geography is not the physical landscape of the Earth, it is not possible to discuss human geography without going into the physical landscape with which human activities are being played out and environmental geography which is an important link between the two.

Human geography is both methodologically and theoretically diverse, including feminist, marxist, post-structural approaches, among others, and using both qualitative methods (such as ethnographic and interviews) and quantitative methods (such as survey research, statistical analysis and model building). In the history of geography, geographers have often recorded and described features of the Earth that might now be considered the remit of human, rather than physical, geographers. For example Hecataeus of Miletus, a geographer and historian in ancient Greece, described inhabitants of the ancient world as well as physical features. It was not until the 18th and 19th Centuries, however, that geography was recognised as a discrete academic discipline. The Royal Geographical Society was founded in England in 1830, although the United Kingdom did not get its first full Chair of geography until 1917. The first real geographical intellect to emerge in United Kingdom geography was Halford John Mackinder, appointed reader at Oxford University in 1887. The National Geographic Society was founded in the USA in 1888 and began publication of the *National Geographic* magazine which became and continues to be a great popularizer of geographic information. The society has long supported geographic research and education.

One of the first examples of geographic methods being used for purposes other than to describe and theorise the physical properties of the earth is John Snow's map of the 1854 Broad Street cholera outbreak. Though a physician and a pioneer of epidemiology, the map is probably one of the earliest examples of Health geography.

However, the now fairly distinct differences between the subfields of physical and human geography developed at a later date. This connection between both physical and human properties of geography is most apparent in the theory of Environmental determinism, made popular in the 19th Century by Carl Ritter and others, and with close links to evolutionary biology of the time. Environmental determinism is the theory that a people's physical, mental and moral habits are directly due to the influence of their natural environment. However, by the mid 19th Century, environmental determinism was under attack for lacking methodological rigour associated with modern science, and later as serving to justify racism and imperialism

A similar concern with both human and physical aspects is apparent in the later Regional geography, during the later 19th and first half of the 20th Centuries. The goal of regional geography, through regionalization, was to delineate space into regions and then understand and describe the unique characteristics of each region, in both human and physical aspects. With links to possibilism and cultural ecology, some of the same notions of causal effect of the environment on society and culture, as with environmental determinism remained.

By the 1950s, however, the quantitative revolution lead to strong criticism of regional geography. Due to a perceived lack of scientific rigour in and overly descriptive nature of the discipline, and a continued separation of geography from geology and the two subfields of physical and human geography, geographers in the mid 19th Century began to apply statistical and mathematical model methods to solving spatial problems. Much of the development during the quantitative revolution is now apparent in the use of Geographic information systems, and the use of statistics, spatial modelling and positivist approaches is still important to many branches of human geography. Well-known geographers from this period are Fred K. Schaefer, Waldo Tobler,

William Garrison, Peter Haggett, Richard J. Chorley, William Bunge, and Torsten Hagerstrand.

From the 1970s a number of critiques of the positivism now associated with geography emerged. Known under the term critical geography this signalled another turning point in the discipline. Behavioural geography emerged for some time as a means to understand how people made perceived spaces and places, and made locational decisions.

More influentially, radical geography emerged in the 1970s and 1980s, drawing heavily on Marxist theory and techniques, and is associated with geographers such as David Harvey and Richard Peet. Seeking to say something 'meaningful' about the problems recognised through quantitative methods, to provide explanations rather than descriptions, to put forward alternatives and solutions and to be politically engaged, rather than the detachment associated with positivist methods. (The detachment and objectivity of the quantitative revolution was itself critiqued by radical geographers as being a tool of capital). Radical geography and the links to Marxism and related theories remain an important part of contemporary human geography) Critical geography also saw the introduction of humanistic geography, associated with the work of Yi-Fu Tuan, which, though similar to behavioural geography, pushed for a much more qualitative approach in methodology. The changes under critical geography have lead to contemporary approaches in the discipline such as Feminist geography, New cultural geography, and the engagement with postmodern and post structural theories and philosophies.

Fields of Human Geography

The main fields of study in human geography focus around the core fields of:

Culture

Cultural geography is the study of cultural products and norms and their variation across and relations to spaces and places. It focuses on describing and analysing the ways language, religion, economy, government, and other cultural phenomena vary or remain constant from one place to another and on explaining how humans function spatially.

- Subfields include: Children's geographies, Animal geographies, Language geography, Sexuality and Space and Religion geography.

Development

Development Geography is the study of the Earth's geography with reference to the Standard of living and the Quality of life of its human inhabitants, study of the location, distribution and spatial organization of economic activities across the Earth. The subject matter investigated is strongly influenced by the researcher's methodological approach.

Economic

Economic geography examines relationships between human economic systems, states, and other actors, and the biophysical environment.

Health

Health geography is the application of geographical information, perspectives, and methods to the study of health, disease, and health care.

Historical

Historical Geography is the study of the human, physical, fictional, theoretical, and "real" geographies of the past. Historical geography studies a wide variety of issues and topics. A common theme is the study of the geographies of the past and how a place or region changes through time. Many historical geographers study geographical patterns through time, including how people have interacted with their environment, and created the cultural landscape.

Political

Political geography is concerned with the study of both the spatially uneven outcomes of political processes and the ways in which political processes are themselves affected by spatial structures.

Population

Population geography is the study of the ways in which spatial variations in the distribution, composition, migration, and growth of populations are related to the nature of places.

Tourism

Tourism geography is the study of travel and tourism as an industry, as a human activity, and especially as a place-based experience.

Urban

Urban geography is the study of urban areas with specific regards to spatial and relational aspects and theories. That is the study of areas which have a high concentration of buildings and infrastructure. These are areas where the majority of economic activities are in the secondary sector and tertiary sectors. They probably have a high population density.

Cultural Geography

Cultural geography is a sub-field within human geography. Cultural geography is the study of cultural products and norms and their variations across and relations to spaces and places. It focuses on describing and analysing the ways language, religion, economy, government and other cultural phenomena vary or remain constant, from one place to another and on explaining how humans function spatially.

Areas of Study

The areas of study of cultural geography are very broad. Among many applicable topics within the field of study are:

- Globalization has been theorised as an explanation for cultural convergence.
- Westernization or other similar processes such as modernization, americanization, is lamization and others.
- Theories of cultural hegemony or cultural assimilation via cultural imperialism.
- Cultural a real differentiation, as a study of differences in way of life encompassing ideas, attitudes, languages, practices, institutions and structures of power and whole range of cultural practices in geographical areas.
- Study of cultural landscapes.
- Other topics include spirit of place, colonialism, post-colonialism, internationalism, immigration, emigration and ecotourism.

History

Though the first traces of the study of different nations and cultures on Earth can be dated back to ancient geographers such as Ptolemy or Strabo, cultural geography as academic study firstly emerged as an alternative to the environmental determinist theories of the early Twentieth century, which had believed that people and societies are controlled by the environment in which they develop. Rather than studying pre-determined regions based upon environmental classifications, cultural geography became interested in cultural landscapes. This was led by Carl O. Sauer (called the father of cultural geography), at the University of California, Berkeley. As a result, cultural geography was long dominated by American writers.

Sauer defined the landscape as the defining unit of geographic study. He saw that cultures and societies both developed out of their landscape, but also shaped them too. This interaction between the 'natural' landscape and humans creates the 'cultural landscape'. Sauer's work was highly qualitative and descriptive and was surpassed in the 1930s by the regional geography of Richard Hartshorne, followed by the quantitative revolution. Cultural geography was generally sidelined, though writers such as David Lowenthal continued to work on the concept of landscape. In the 1970s, the critique of positivism in geography caused geographers to look beyond the quantitative geography for its ideas. One of these re-assessed areas was also cultural geography.

New Cultural Geography

Since the 1980s, a "new cultural geography" has emerged, drawing on a diverse set of theoretical traditions, including Marxist political-economic models, feminist theory, post-colonial theory, post-structuralism and psychoanalysis.

Drawing particularly from the theories of Michel Foucault and performativity in western academia, and the more diverse influences of postcolonial theory, there has been a concerted effort to deconstruct the cultural in order to make apparent the various power relations. A particular area of interest is that of identity politics and construction of identity.

Examples of areas of study include:

- Feminist geography
- Children's geographies
- Some parts of Tourism geography
- Behavioural geography
- Sexuality and space
- Some more recent developments in Political geography.

Some within the 'new cultural geography' have turned their attention to critiquing some of its ideas, seeing its views on identity and space as static. It has followed the critiques of Foucault made by other 'post structuralist' theorists such as Michel de Certeau and Gilles Deleuze. In this area, non-representational geography and population mobility research have dominated. Others have attempted to incorporate these critiques back into the new cultural geography.

Children's Geographies

Children's geographies is an area of study in human geography, studying the places and spaces of children's lives. Children's geographies is that branch of human geography which deals with the study of places and spaces of children's lives, characterised experientially, politically and ethically.

Ever since the cultural turn in geography, there has been recognition that society is not homogenous but heterogeneous. It is characterized by diversity, differences and subjectivities. While feminist geographers had been able to strengthen the need for examination of gender, class and race as issues affecting women, 'children' as an umbrella term encompassing children, teenagers, youths and young people, which are still relatively missing as a 'frame of reference' in the complexities of 'geographies'.

In the act of theorizing children and their geographies, the ways of doing research and the assumed ontological realities often "frame 'children' and 'adults' in ways that impose a bipolar, hierarchical, and developmental model". This reproduces and enforces the hegemony of adult-centered discourses of children within knowledge production. Children's geographies has developed in academic human geography since the beginning of the 1990s, although there were notable studies in the area before that date. The earliest work done on children's geographies largely can be traced to William Bunge's work on spatial oppression of children in Detroit and Toronto where children are deemed as the ones who suffer the most under an oppressing adult framework of social, cultural and political forces controlling the urban built environment.

This development emerged from the realisation that previously human geography had largely ignored the everyday lives of children, who (obviously) form a significant section of society, and who have specific needs and capacities, and who may experience the world in very different ways. Thus children's geographies can in part be seen in parallel to an interest in gender in geography and feminist geography in so much as their starting points were the gender blindness of mainstream academic geography.

Children's geographies rests on the idea that children as a social group share certain characteristics which are experientially, politically and ethically significant and which are worthy of study. The pluralisation in the title is intended to imply that children's lives will be markedly different in differing times and places and in differing circumstances such as gender, family, and class. The current developments in children's geographies are attempting to link the frame of analysing children's geographies to one that requires multiple perspectives and the willingness to acknowledge the 'multiplicity' of their geographies. Children's geographies is sometimes coupled with, and yet distinguished from the geographies of childhood. The former has an interest in the everyday lives of children; the latter has an interest in how (adult) society conceives of the very idea of childhood and how this impinges on children's lives in many ways. This includes imaginations about the nature of children and the related (spatial) implications.

There are a whole range of foci with children's geographies including children and the city, children and the countryside, children and technology, children and nature, children and globalisation, methodologies of researching children's worlds and the ethics of doing so; see the otherness of childhood. There is now a Journal of Children's Geographies which will give readers a good idea of the growing range of issues, theories and methodologies of this developing and vibrant sub-discipline.

Animal Geographies

Animal geographies is an area of study in geography, studying the spaces and places occupied by animals in human culture.

An interest in animal geographies emerged in human geography in the mid 1990s. This was marked by a special edition of the journal *Environment and Planning D: Society and Space* in 1995 and a book by Jennifer Wolch and Jody Emel called *Animal Geographies: place, politics and identity in the nature-culture borderlands* published in 1998.

This movement was prompted by the basic fact that social life and space is heavily populated by animals of many differing kinds and in many differing ways (e.g. farm animals, pets, wild animals in the city). It was also prompted by ecofeminist and other environmentalist viewpoints on nature-society relations (including questions of animal welfare and rights).

This sub-discipline within human geography quickly developed, another landmark being the book *Animal Spaces, Beastly Places: New Geographies of Human-Animal Relations.*

Papers regularly come out in a number of geographical journals and in journals such as *Society and Animals*. Animal geographies is now part of a wider interest of non-human or 'more-than-human' geographies which pays close attention not only to animals but all the things, living and non-living, that help to make up the everyday 'social' world and its spaces and places.

Language Geography

Language geography is the branch of human geography that studies the geographic distribution of language or its constituent elements. There are two principal fields of study within the geography of language: the "geography of languages", which deals with the distribution through history and space of languages, and "linguistic geography", which deals with regional linguistic variations within languages. Various other terms and subdisciplines have been suggested, including; a division within the examination of linguistic geography separating the studies of change over time and space; 'geolinguistics', a study within the geography of language concerned with 'the analysis of the distribution patterns and spatial structures of languages in contact', but none have gained much currency.

Many studies have researched the effect of 'language contact', as the languages or dialects of peoples have interacted. This territorial expansion of language groups has usually resulted in the overlaying of languages upon existing speech areas, rather than the replacement of one language by another. An example could be sought in the Norman Conquest of England, where Old French became the language of the aristocracy, and Middle English remained the language of the majority of the population.

Linguistic geography, as a field, is dominated by linguists rather than geographers. Charles Withers describes the difference as resulting from a focus on "elements of language, and only then with their geographical or social variation, as opposed to investigation of the processes making for change in the extent of language areas." To quote Trudgill, "linguistic geography has been geographical only in the sense that it has been concerned with the spatial distribution of linguistic phenomena." In recent times greater emphasis has been laid upon explanation rather than description of the patterns of linguistic change. The move has paralleled similar concerns in geography and language studies. These studies have paid attention to the social use of language, and to variations in dialect within languages in regard to social class or occupation. Regarding such variations, lexicographer Robert Burchfield notes that their nature "is a matter of perpetual discussion and disagreement". As an example, he notes that "most professional linguistic scholars regard it as axiomatic that all varieties of English have a sufficiently large vocabulary for the expression of all the distinctions that are important in the society using it."

He contrasts this with the view of the historian Professor John Vincent, who regards such a view as "a nasty little orthodoxy among the educational and linguistic establishment. However badly you need standard English, you will have the merits of non-standard English waved at you. The more extravagantly your disadvantages will be lauded as 'entirely adequate for the needs of their speakers', to cite the author of *Sociolinguistics.* It may sound like a radical cry to support pidgin, patois, or dialect, but translated into social terms, it looks more like a ploy to keep Them (whoever Them may be) out of the middle-class suburbs."

— *John Vincent, The Times*

Burchfield concludes that "Resolution of such opposite views is not possible", though the "future of dialect studies and the study of class-marked distinctions are likely to be of considerable interest to everyone".

In England, linguistic geography has traditionally focussed upon rural English, rather than urban English. A common production of linguistic invesigators of dialects is the shaded and dotted map showing where one linguistic feature ends and another begins or overlaps. Various compilations of these maps for England have been issued over the years, including Joseph Wright's *English Dialect Dictionary* (1896-1905), the *Survey of English Dialects* (1962-8), and *The Linguistic Atlas of England* (1978).

Religion and Geography

Religion and geography is the study of the impact of geography, i.e. place and space, on religious belief.

Another aspect of the relationship between religion and geography is *religious geography*, in which geographical ideas are influenced by religion, such as early map-making, and the *biblical geography* that developed in the 16th century to identify places from the Bible.

Major Religious Groups

The world's principal religions and spiritual traditions may be classified into a small number of major groups, although this is by no means a uniform practice. This theory began in the 18th century with the goal of recognizing the relative levels of civility in non-European societies. However, it quickly transformed into a subset of the universalist belief that all religious figures teach of a single, cross-cultural truth. For a more comprehensive list of religions and an outline of some of their basic relationships, please see the article list of religions.

History of Religious Categories

In world cultures, there have traditionally been many different groupings of religious belief. In Indian culture, different religious philosophies were traditionally respected as academic differences in pursuit of the same truth. In Islam, the Qur'an mentions three different categories: Muslims, the People of the Book, and idol worshipers. To some extent these theories of religiousness are still prevalent today. However, the most common classification today was birthed out of Western Christianity.

Initially, Christians had a simple dichotomy of world beliefs: Christian civility versus foreign heresy or barbarity. In the eighteenth century, "heresy" was clarified to mean Judaism and Islam; along with outright paganism, this created a fourfold classification which spawned such works as John Toland's *Nazarenus, or Jewish, Gentile, and Mahometan Christianity,* which represented the three Abrahamic traditions as different "nations" or sects within *religion* itself, the true monotheism. At the turn of the 18th century, in between 1780 and 1810, the language dramatically changed: instead of "religion" being synonymous with spirituality, authors began using the plural, "religions", to refer to both Christianity and other forms of worship. This new definition was described as follows by Daniel Defoe: "Religion is properly the Worship given to God, but 'tis also applied to the Worship of Idols and false Deities."

In 1838, the four-way division of Christianity, Judaism, "Mahommedanism" and Paganism was multiplied considerably by Josiah Conder's *Analytical and Comparative View of All Religions Now Extant among Mankind.* Conder's work still adheres to the four-way classification, but in his eye for detail he puts together much historical work to create something resembling our modern Western image: he includes Druze, Yezidis, Mandeans, and Elamites under a list of possibly monotheistic groups, and under the final category, of "polytheism and pantheism", he lists Zoroastrianism, "Vedas, Puranas, Tantras, Reformed sects" of India as well as "Brahminical idolatry", Buddhism, Jainism, Sikhism, Lamaism, "religion of China and Japan", and "illiterate superstitions".

Even through the late nineteenth century, it was common to view these "pagan" sects as dead traditions which preceded Christianity, the final, complete word of God. This in no way reflected the reality of religious experience: Christians supposed these traditions to have maintained themselves in an unchanging state since whenever they were "invented", but actually all traditions survived in the words and deeds of people, some of whom could make radical new inventions without needing to create a new sect.

The biggest problem in this approach was the existence of Islam, a religion which had been "founded" after Christianity, and which had been experienced by Christians as intellectual and material prosperity. By the nineteenth century, however, it was possible to dismiss Islam as a revelation of "the letter, which killeth", given to savage desert nomads. In this context, the term "world religion" referred only to Christianity, which Europeans considered uniquely posed to civilize the world.

The modern meaning of the phrase "world religion" began with the 1893 Parliament of the World's Religions in Chicago, Illinois. This event was sharply criticized by European Orientalists up until the 1960s as "unscientific", because it allowed religious leaders to speak for themselves instead of bowing to the superior knowledge of the Western academic. As a result its approach to "world religion" was not taken seriously in the scholarly world for some time. Nevertheless, the Parliament spurred the creation of a dozen privately funded lectures with the intent of informing people of the diversity of religious experience: these lectures funded researchers such as William James, D.T. Suzuki, and Alan Watts.

In the latter half of the 20th century, the category of "world religion" fell into serious question, especially for drawing parallels between vastly different cultures, and thereby creating an arbitrary separation between the religious and the secular. Even history professors have now taken note of these complications and advise against teaching "world religions" in schools.

Western Classification

Religious traditions fall into super-groups in comparative religion, arranged by historical origin and mutual influence. Abrahamic religions originate in the Middle East, Indian religions in India and Far Eastern religions in East Asia. Another group with supra-regional influence are African diasporic religions, which have their origins in Central and West Africa.

- Abrahamic religions are the largest group, and these consist mainly of Christianity, Islam, Judaism and Bahai Faith. They are named for the patriarch Abraham, and are unified by the practice of monotheism. Today, around 3.4 billion people are followers of Abrahamic religions and are spread widely around the world apart from the regions around South-East Asia. Several Abrahamic organizations are vigorous proselytizers.

- Indian religions originated in Greater India and tend to share a number of key concepts, such as dharma and karma. They are of the most influence across the Indian subcontinent, East Asia, South East Asia, as well as isolated parts of Russia. The main Indian religions are Hinduism, Jainism, Buddhism and Sikhism. Indian religions mutually influenced each other. Sikhism was also influenced by the Abrahamic tradition of Sufism.
- East Asian religions consist of several East Asian religions which make use of the concept of *Tao* (in Chinese) or *Do* (in Japanese or Korean), namely Taoism and Confucianism, both of which are asserted by some scholars to be non-religious in nature.
- African diasporic religions practiced in the Americas, imported as a result of the Atlantic slave trade of the 16th to 18th centuries, building of traditional religions of Central and West Africa.
- Indigenous tribal religions, formerly found on every continent, now marginalized by the major organized faiths, but persisting as undercurrents of folk religion. Includes African traditional religions, Asian Shamanism, Native American religions, Austronesian and Australian Aboriginal traditions, Chinese folk religion, and postwar Shinto. Under more traditional listings, this has been referred to as "Paganism" along with historical polytheism.
- Iranic religions (not listed below due to overlaps) originated in Iran and include Zoroastrianism, Yazdanism, Ahl-e Haqq and historical traditions of Gnosticism (Mandaeanism, Manichaeism). It has significant overlaps with Abrahamic traditions, e.g. in Sufism and in recent movements such as Babism and the Bahai Faith.
- New religious movement is the term applied to any religious faith which has emerged since the 19th century, often syncretizing, re-interpreting or reviving aspects of older traditions: Hindu revivalism, Ayyavazhi, Pentecostalism, polytheistic reconstructionism, and so forth.

Religious Demographics

One way to define a major religion is by the number of current adherents. The population numbers by religion are computed by a combination of census reports and population surveys (in countries where religion data is not collected in census, for example USA or France), but results can vary widely depending on the way questions are phrased, the definitions of religion used and the bias of the agencies or organizations conducting the survey. Informal or unorganized religions are especially difficult to count.

There is no consensus among researchers as to the best methodology for determining the religiosity profile of the world's population. A number of fundamental aspects are unresolved:

- Whether to count "historically predominant religious culture[s]"
- Whether to count only those who actively "practice" a particular religion
- Whether to count based on a concept of "adherence"
- Whether to count only those who expressly self-identify with a particular denomination
- Whether to count only adults, or to include children as well.
- Whether to rely only on official government-provided statistics
- Whether to use multiple sources and ranges or single best source(s).

Largest Religions or Belief Systems by Number of Adherents

However, religious philosophy is not always the determining factor in local practice. Please note that heterodox movements as adherents to their larger philosophical category, although this may be disputed by others within that category. For example, Cao Dai is listed because it claims to be a separate category from Buddhism, while Hoa Hao is not, even though they are similar new religious movements.

The population numbers below are computed by a combination of census reports, random surveys (in countries where religion data is not collected in census, for example USA or France), and self-reported attendance numbers, but results can vary widely depending on the way questions are phrased, the definitions of religion used and the bias of the agencies or organizations conducting the survey. Informal or unorganized religions are especially difficult to count. Some organizations may wildly inflate their numbers.

Development Geography

Development geography is the study of the earth's geography with reference to the standard of living and quality of life of its human inhabitants. In this context, development is a process of change that affects people's lives. It may involve an improvement in the quality of life as perceived by the people undergoing change. However, development is not always a positive process. Gunder Frank commented on the global economic forces that lead to the development of underdevelopment. This is covered in his dependency theory.

In development geography, geographers study spatial patterns in development. They try to find by what characteristics they can measure development by looking at economic, political and social factors. They seek to understand both the geographical causes and consequences of varying development. Studies compare More Economically Developed Countries (MEDCs) with Less Economically Developed Countries (LEDCs). Additionally variations within countries are looked at such as the differences between northern and southern Italy, the Mezzogiorno.

Within development geography, sustainable development is also studied in an attempt to understand how to meet the needs of the present without compromising the needs of future generations to meet their own needs.

Quantitative Indicators

Quantitative indicators are numerical indications of development:

- Economic indicators include GNP (Gross National Product) per capita, unemployment rates, energy consumption and percentage of GNP in primary industries. Of these, GNP per capita is the most used as it measures the value of all the goods and services produced in a country, excluding those produced by foreign companies, hence measuring the economic and industrial development of the country. However, using GNP per capita also has many problems.
    - It does not take into account the distribution of the money which can often be extremely unequal as in the UAE where oil money has been collected by a rich elite and has not flowed to the bulk of the country.
    - GNP does not measure whether the money produced is actually improving people's lives and this is important because in many MEDCs where there are large increases in wealth over time but only small increases in happiness.
    - The figure rarely takes into account the unofficial economy, which includes subsistence agriculture and cash-in-hand or unpaid work, which is often substantial in LEDCs. In LEDCs it is often too expensive to accurately collect this data and some governments intentionally or unintentionally release inaccurate figures.

- o The figure is usually given in US dollars which due to changing currency exchange rates can distort the money's true street value so it is often converted using purchasing power parity (PPP) in which the actual comparative purchasing power of the money in the country is calculated.
- Social indications include access to clean water and sanitation (which indicate the level of infrastructure developed in the country) and adult literacy rate, measuring the resources the government has to meet the needs of the people.
- Demographic indicators include the birth rate, death rate and fertility rate, which indicate the level of industrialization.
  - o Health indicators (a sub-factor of demographic indicators) include nutrition (calories per day, calories from protein, percentage of population with malnutrition), infant mortality and population per doctor, which indicate the availability of healthcare and sanitation facilities in a country.
- Environmental indications include how much a country does for the environment.

Composite Indicators

Composite or qualitative indicators combine several quantitative indicators into one figure and generally provide a more balanced view of a country. Usually they include one economic, one social and one demographic indicator.

- The HDI (Human Development Index) is now the most widely used composite indicator. A number is calculated between 0 and 1 taking into account the most important measures: GNP per capita, the adult literacy rate, the school enrollment rate and life expectancy. It was started by the United Nations in 1990 to replace GNP as a more accurate way of measuring development. A HDI between 1 and 0.8 is considered high, 0.8 and 0.6 is considered medium and 0.6 to 0.4 is considered low.

Data Example

HDI Rank	*Country*	*GDP Per Capita (PPP US$) 2008*	*Human Development index (HDI) value 2006*
4	Australia	35,677	0.965
70	Brazil	10,296	0.807
151	Zimbabwe	188	0.513

- Other composite measures include the PQLI (Physical Quality of Life Index) which was a precursor to the HDI which used infant mortality rate instead of GNP per capita and rated countries from 0 to 100. It was calculated by assigning each country a score of 0 to 100 for each indicator compared with other countries in the world. The average of these three numbers makes the PQLI of a country.

- The HPI (Human Poverty Index) is used to calculate the percentage of people in a country who live in relative poverty. In order to better differentiate the number of people in abnormally poor living conditions the HPI-1 is used in developing countries, and the HPI-2 is used in developed countries. The HPI-1 is calculated based on the percentage of people not expected to survive to 40, the adult illiteracy rate, the percentage of people without access to safe water, health services and the percentage of children under 5 who are underweight. The HPI-2 is calculated based on the percentage of people who do not survive to 60, the adult functional illiteracy rate and the percentage of people living below 50% of median personal disposable income.
- The GDI (Gender-related Development Index) measures gender equality in a country in terms of life expectancy, literacy rates, school attendance and income.

Qualitative Indicators

Qualitative indicators include descriptions of living conditions and people's quality of life. They are useful in analysing features that are not easily calculated or measured in numbers such as freedom, corruption or security, which are mainly non-material benefits.

Geographic Variations in Development

The updated view of the north-south divide. Blue includes G8 nations, developed/first world nations, and Europe

There is a considerable spatial variation in development rates.

Global wealth also increased in material terms, and during the period 1947 to 2000, average per capita incomes tripled as global GDP increased almost tenfold (from $US3 trillion to $US30 trillion)... Over 25% of the 4.5 billion people in LEDCs still have life expectancies below 40 years. More than 80 countries have a lower annual per capita income in 2000 than they did in 1990. The average income in the world's five richest countries is 74 times the level in the world's poorest five, the widest it has ever been. Nearly 1.3 billion people have no access to clean water. About 840 million people are malnourished.—Stephen Codrington

The most famous pattern in development is the North-South divide.

The North-South divide is the divide which separates the rich North or the developed world, from the poor South. This line of division is not as straightforward as it sounds and splits the globe into two main parts. It is also known as the Brandt Line.

The “North” in this divide is regarded as being North America, Europe, Russia, South Korea, Japan, Australia, New Zealand and the like. The countries within this area are generally the more economically developed. The “South” therefore encompasses the remainder of the Southern Hemisphere, mostly consisting of LEDCs. Another possible dividing line is the Tropic of Cancer with the exceptions of Australia and New Zealand.

It is critical to understand that the status of countries is far from static and the pattern is likely to become distorted with the fast development of certain southern countries, many of them NICs (Newly Industrialised Countries) including India, Thailand, Brazil, Malaysia, Mexico and others.

These countries are experiencing sustained fast development on the back of growing manufacturing industries and exports.

Most countries are experiencing significant increases in wealth and standard of living. However there are unfortunate exceptions to this rule.

Noticeably some of the former Soviet Union countries has experienced major disruption of industry in the transition to a market economy. Many African nations have recently experienced reduced GNPs due to wars and the AIDS epidemic, including Angola, Congo, Sierra Leone and others.

Arab oil producers rely very heavily on oil exports to support their GDPs so any reduction in oil’s market price can lead to rapid decreases in GNP. Countries which rely on only a few exports for much of their income are very vulnerable to changes in the market value of those commodities and are often derogatively called banana republics. Many developing countries do rely on exports of a few primary goods for a large amount of their income (coffee and timber for example), and this can create havoc when the value of these commodities drops, leaving these countries with no way to pay off their debts.

Within countries the pattern is that wealth is more concentrated around urban areas than rural areas. Wealth also tends towards areas with natural resources or in areas that are involved in tertiary (service) industries and trade. This leads to a gathering of wealth around mines and monetary centres such as New York, London and Tokyo.

Aid

MEDCs (More Economically Developed Countries) can give aid to LEDCs (Less Economically Developed Countries). There are several types of aid:

- Governmental (Bilateral) aid
- International Organizational (multilateral) aid
- Voluntary aid
- Short-term/emergency aid
- Long-term/sustainable aid.

Aid can be given in several ways. Through money, materials, or skilled and learned people (e.g. teachers).

Aid has advantages. Mostly short-term or emergency aid help people in LEDCs to survive a natural (earthquake, tsunami, volcano eruption etc.) or human (civil war etc.) disaster. Aid helps make the recipient country (the country that receives aid) get more developed.

However, aid also has disadvantages. Often aid does not even reach the poorest people. Often money gained from aid is used up to make infrastructures (bridges, roads etc.), which only the rich can use. Also, the recipient country gets more dependant of aid from a donor country (the country giving aid).

2

Man and Environment

Environment may be defined as the surrounding conditions—physical or non-physical, that influence and modify the life of an individual or a community. It influences the habits and modes of life of man and his degree of civilization. Some groups of people in particular environmental conditions have profound possibilities for economic activities, while others have to work in either limited or adverse conditions. The particular environmental conditions of U.K., Germany and Japan have made them industrially and economically very advanced. The fertile river valley of China and India together with monsoon type of climate have made these countries agriculturally rich. The people of Savannah are, as a rule, the great hunters of the world. The environmental conditions are favourable for wild animals—carnivores and herbivores. The temperate grasslands of the world are the granaries of the world. All the principal wheat-lands of the world are associated with these grasslands. The ever hot, ever wet climates have made the Amazonian people indolent and uncenterprising. On the other hand, the mild equable climate of Western Europe has produced energetic, active and most enterprising people.

We must not, however, forget that the same type of environment may not necessarily produce a similar mode of living. Some groups of people may derive more benefit from their environment, while others living in similar conditions may not be well advanced in the field of economic activity. The quality and the culture of the people are the dominating factors in the utilization of economic resources in different environmental conditions. The economic development of Australia and South Africa was made not by the local people but by the people of Western Europe, who were endowed with high culture and superior quality. These people are scientifically well advanced. They could make the best utilization of the economic resources available in these regions. The Congo and the Amazon basins, and Indonesia have the same type of environmental conditions. But owing to the fact that the people of Indonesia are more advanced than the other two regions, the country is economically and politically more advanced and developed. Of course, owing to the insular situation of Indonesia, white settlement was possible. The contribution of the white people in developing the region was no less important. So, though the mode of life in any region is influenced by the factors of environment, it is the quality and culture of the people that matter much in carrying on their economic activities.

The factors of environment can broadly be classified into two groups—physical and non-physical.

The principal physical factors are (a) Geographical location, (b) coastline, (c) Topography, (d) inland water bodies, (e) climate, (f) soil, (g) animal, (h) vegetation, (i) minerals, etc.

The principal non-physical factors are (a) Population, (b) Political and Social organization.

Geographical Location

The geographical location of a country in relation to the surface of the globe determines to a great extent, the industrial and commercial development.

The location of a country can be studied from the following points of view:—

The location with reference to land and sea: The location with reference to land and sea may be Insular, Littoral or marginal, Isthmian, Peninsular and Continental.

When a country is surrounded by the sea, the location is called *insular*. This provides access to the sea on all sides of the country. All the ocean routes of the world become accessible to the country. Sea-winds have moderating effect on climate. This type of land produces sea-faring nations. People become hardy and skilful and commercially and industrially advanced. Britain and Japan have insular positions. They have established their trade and commerce through ocean routes with the different countries of the world. Moreover, the defence of a country having a sea-frontier becomes comparatively easier.

Littoral or marginal location means that a considerable portion of a country is located by the side of the sea. The area is influenced by sea winds and can carry on trade along the coast by sea route. France, Norway, Sweden and Spain enjoy this advantage.

An isthmus is a narrow strip of land that connects two larger land masses. The isthmus of Panama is an important example. This connects the two great continents, namely, the North America and the South America. By cutting this isthmus two mighty oceans, namely, the Atlantic and the Pacific have been linked up. This has offered a great possibility of carrying on trade between the western and the eastern coasts of the U.S.A. and the western and northern coastal countries of South America through ocean route.

Peninsular location means that a country or a region is surrounded by the sea on three sides. Southern India, Malyasia, Italy, etc., have this type of location. These countries can carry on trade and commerce both by sea and land-routes. Sea-winds also have influence on their climate. The location of a country is said to be *continental* when it is located far away from the sea surrounded by landmasses. The climate of this region becomes extreme. Summer temperature becomes high and winter temperature becomes low. Sea-borne trade becomes difficult. Contact with foreign countries becomes limited. Defence of land frontier becomes a great problem. Afghanistan, Bolivia, Poland etc., offer good examples of continental location. A country like Nepal has to depend on India to carry on trade with other countries.

Location with reference to latitude: On the variation of latitude, a country may be located in tropical, sub-tropical, temperate or polar regions. Tropical and sub-tropical countries like India, Indonesia, etc., have hot and humid climate. Agricul-turally these countries are rich. The principal agricultural crops are rice, jute, cotton, sugar-cane, rubber, tea, etc. Temperate countries have moderate climate. The people are energetic. Industrially these countries are rich. They produce temperate crops like wheat, barley, sugar-beet, etc. Mechanization in agriculture is important. Sea-fishing near the sea-coasts is an important occupation. Countries like the U.K., Germany, France, Japan, Norway offer good examples. In the polar regions of Canada, Soviet Union, and Greenland economic activity is limited. Seal-fishing and special type of animal hunting are the principal occupations. Eskimos of Canada are important to note.

Location in relation to other Advanced countries: If a country is situated near an industrially and commercially developed country, its economy may be influenced by that country. Western European countries for their industrial and commercial development owe much to the U.K. Progress of newly formed Bangladesh depends much on its relation with India, a more advanced and progressive country, from the point of view of agriculture, industry, trade and commerce. A country remains backward when it fails to come in contact with outside world.

Altitude also a *modifying* effect on climate. Quito is situated on the equator. It should have ever hot, ever wet climate. But because it is situated at a great altitude, its temperature has been modified. Hence it is the land of "*eternal spring*."

Nature of Coastline

The coastline is a line which indicates the line of separation between land and sea. Therefore, all countries with littoral, peninsular or insular position must have sea-coasts. The coastline has a great contribution in developing ports and harbours. The coastline of a country may be indented (broken) and unindented (straight). The coastline may also be high or low.

The *indented* or *broken* coast has great commercial importance. The broken coast helps intrusion of sea into the land. This becomes ideal for the construction of natural ports and harbours, through which sea-borne trade is carried on. Shipping yard can be developed for accommodating and building ships. Through these ports sea-fishing is developed. The port of Visakhapatnam, Cardiff, etc., are ideal sea-ports on broken coasts. Broken coasts of Greece and Norway have made them a sea-faring nations. Broken coasts bring interior landmass near to the sea. The country with broken coasts has larger numbers of natural ports and harbours. Great Britain has many natural ports and harbours.

The *unindented* or *straight* coast : A coastline without a cut into the land by sea is called unindented coast. Such a coast offers little possibility of developing natural port and harbour. The coastline of Africa is not so indented. India with a coastline of about 6083 km, —because of unindented coast for a long distance—has only a small number of ideal ports and harbours like Bombay, Goa, Cochin, and Visakhapatnam. Madras, an artificial port, entails a great cost for its maintenance.

The coastline may be *high* or *low*. The coastline of Africa is locked by mountains. Access to the sea becomes difficult. Adjacent seas are shallow. Development of ports and harbours is difficult. On the other hand, the coastline of North-Western European countries is low as well as indented, thus offering great facilities for the development of natural port and harbour.

Topography

Topography is the nature of land surface and is a reflection of the Geological structure. Topography of the land surface can be divided into three features; namely, *mountains, plateaus,* and *plains.* These three landforms have great influence on the economic development of a country. Plain lands are more advanced that the mountanious and plateau areas.

Influence of Mountains: Though in general mountains offer a great resistance in developing a region economically, yet many facilities man can derive from the mountains for economic development.

Facilities Offered by the Mountains: (i) Mountains act as natural boundary, between two political units. Mountainous frontier offers impediments to easy access by the enemy. The northern and eastern frontiers of India, locked by the mighty Himalayas and its offshoots are more safe. (ii) Mountains are the sources of great rivers. The mighty rivers like the Ganges, the Brahmaputra and the Indus are fed by Himalayan glaciers and they are perennial. They have profound influence on the economic life of the regions through which they pass. Mountain torrents are highly suitable for the generation of hydroelectricity—the cheapest source of power in modern industry. These hydel powers play the most vital part on the economic life of Japan. India's economy is also highly influenced by the hydel power derived from mountain torrents. (iii) Mountains influence the climate of a country. The South-West monsoon winds striking against the Himalayas supply heavy rain to the different parts of India. On the other hand, the Himalayas protect India from the cold wind of the North. (iv) Mountain at its different heights and slopes contains different types of vegetations from hardwood forest to softwood forest—which have great commercial use. The Himalayas and the Rockies are full of such lucrative forests. (v) Mountain slopes are very ideal for the production of a special variety of agricultural crops such as tea, rubber, coffee, etc. Himalayan slopes are ideal for tea, Malayasian mountain slopes for rubber and Brazilian plateau for coffee. (vi) Various types of economic minerals and rocks are obtained from the mountains. Younger rocks contain oil and coal and harder rocks, contain iron, copper, lead, etc. (vii) High mountain top with beautiful natural scenery and healthy climate offers creation of recreation centres and health resorts. Darjeeling, Simla and Nainital are famous in India. Switzerland with beautiful mountain climate is called the playground of Europe. These health-resorts attract tourists, encourage hotel-keeping. The people of mountainous regions are strong, healthy and brave. The Nepalese are renowned warriors.

Limitations Offered by the Mountains: Mountains in many ways offer great resistance in developing the region. The discontinuous uneven mountain terrain stands in the way of developing communication system. Cost of construction is very high. Movement through mountain terrain is not easy and settlement is difficult. Hence the mountainous region is thinly populated. Landslide, accumulation of huge fost and snow stand in the way of developing the region. Soil erosion on mountain slopes is great. Fertility of soil is easily lost. Agriculture is carried on in restricted areas. In recent times, however, man is trying to use the hill slopes in many ways.

Influence of Plateau: Plateaus have flat surfaces and stand well above the neighbouring lands. The river valleys are deep and steep-sided. The steepness of the plateau edges often makes it difficult to reach the top. From the economic point of view the plateau is superior to mountains but inferior to plains in respect to the possibility for economic development. The plain top of the plateau with greater area can be used for various economic purposes. Development of transport system is easier and except on the plateau slopes, the soil erosion is not so great on the top. Soil is comparatively fertile. Plateau is generally a storehouse of minerals. Western Australian plateau is full of minerals like gold, copper and lead. The swift rivers of the plateau slopes are very ideal for the generation of hydro-electricity. The "fall line" of the Appalachian region of the U.S.A. is an outstanding example of hydro-electric source along the plateau margin. Lava plateau of Western India is a great cotton-growing area.

Influence of Plains: Of the different topographical features, plains play the most vital role in developing the economy of a country. Fifty percent of the world's land surface is occupied by different types of plain and plain is the home of about 90% of the world population. Hence, plain is the most densely populated area with varieties of economic activities.

Plain land offers following facilities for man :—

- The level and more or less even surface of plain with rich fertile soil is ideal for agriculture. Soil erosion is less. The plain is thus ideal for developing irrigation systems with rivers, canals and wells. Some of the great river valleys and deltas of the world, namely, those of the Ganges, the Brahmaputra, the Volga, the Mississippi are great agricultural lands. The plain temperate grasslands of North America and South America are noted for the production of foodgrains and other crops.

- In the plain region development of roads, railways, inland waterways like canals are easier and they together with rivers offer great facilities for transport. The Ganges valley with roads, railways, canals, etc., is well developed from the communication point of view.
- In the plain region industrial, trade and commercial centres, cities and towns are easily built up.

Some desert plains are not suitable for human settlement. But the sandy barren Nile valley of Egypt, Rajasthan area in India have been made fertile by developing irrigation system. In short, the plain region is highly developed from the point of view of agriculture, industry, trade and commerce.

Influence of Island Water Bodies

About 70% of the surface of the earth is occupied by water-bodies. Of these, oceans and rivers are the most important. Inland water-bodies, however, consist of rivers, lakes, canals, ponds, etc. Of these water-bodies, again, rivers and lakes play the most important part in the economic life of man.

Influence of Rivers: Of all the physical factors of environment none has played a more important role than rivers and river valleys in helping man's progress and civilization in many ways. Some of the river valleys of the world, namely, those of the Nile, the Ganges, the Idus, the Tigris and Euphratis, the Hwang Ho, etc., are the cradles of earliest civilization. Most of the river valleys and deltas are thickly populated.

The following benefits are derived from the rivers :

- A river supplies water for drinking and fish for food. The Hooghly is the source of drinking water for the great city of Calcutta. The Ganges, the Brahmaputra, etc. are the sources of fish.
- In the upper course, the river is swift and torrential. In this part, the river is destructive in nature. But the swift flow of water is utilized for the generation of water-power, the cheapest source of power in industry today.
- In the middle and lower courses of the river, by constructing irrigation canals water from the river is carried on to the agricultural land for irrigation purpose. This irrigation system

plays the most important part in agriculture. A network of canals in the Punjab, Rajasthan, U.P., etc., carrying water from the rivers have turned the dry and semi-desert areas into principal agricultural land. The Nile with its canals have turned the desert valley and delta into well-developed agricultural land. Hence Egypt is called the gift of the Nile. Had there be no Nile, Egypt would remain a desert land.

- The rivers along with its canals play the most vital part as inland communication system. Cities and towns have been developed on both banks of the river. Some of the great ports of the world such as London, and Calcutta are situated respectively on the Thames and the Hooghly. A river gives passage to the sea—thus helping sea-borne trade.
- A river brings with it huge amount of silt and deposits them on both banks of the river and on deltas making those regions fertile. Hence the river valleys and deltas are agriculturally rich.
- Some of the rivers act as boundary-line between two political units. The Danube of Europe is an international river crossing the boundaries of Germany, Austria, Czechoslovakia, Yugoslavia, Rumania and Bulgaria.
- Some of the rivers offer recreation facilities such as swimming, boating etc.

But some of the rivers by causing flood, erosion etc., bring about miseries to the life and property of the people of the region. The Damodar of West Bengal and Hwang Ho of China were the sorrows of the respective areas through which they pass. But today by developing multipurpose river valley projects some of these rivers have been tamed. By these projects not only has the flood been controlled but also the following objects have been accomplished. These are : (i) Checking of flood and soil erosion, (ii) irrigation facility, (iii) production of water power, (iv) transport facility, etc. In their turn, these have given rise to various other economic activities.

Another source of inland waterways is some of the *great lakes* of the world. In the U.S.A. and Canada, the five Great Lakes play the most important part as inland waterways. Coal, iron ore, nickel, and various types of agricultural and industrial products are carried on by both the countries through these water bodies.

In France, Germany and Soviet Union the network of canal system plays the most important part as inland waterways.

Influence Lakes and Canals

Influence of Lakes

A lake is a sheet of water surround by land. The water occupies a basin in the earth's surface. Lakes may be of different types on the basis of size, shape, depth, amount of salt present with water and mode of formation.

Lakes also play an important part in influencing the economic activites of man.

Influence of Lakes on Climate: Like ocean lakes also as water body influence the climate of a place. Lake water in summer lowers the temperature of air and increases the temperature in winter month. The five great lakes of North America namely, Superior, Michigan, Huron, Erie and Ontario modifies the temperature of the region and by increasing the moisture of air through evaporation help in increasing the rainfall. Caspean Sea is the biggest salt water lake. The Elburz mountain lies South of it. The winds blowing over the Caspean Sea absorb moisture and these moisture ladden winds when strike against the northern slope of the mountain discharge rain water.

Influence of Lakes as Recreation Centre: Lakes play the most important part as recreation centre. Beautiful scenery attracts tourists from different parts. The lakes of Kashmir are very important in this respect.

Sources of Minerals: The water of the lakes are the sources of minerals and fish. By drying the salty water of lakes *salt,* very useful to man, is obtained. Varieties of fish are obtained from the lake. The Chilka lake of India supplies fish which is sent to different states of India.

Uses of Lakes as Inland Water-ways: Lakes, in some parts of the world, play the most vital part as inland water-route. The five great lakes of North America play the most vital part as means of communication. They have been joined by canals to make the water route perfect. Internal and foreign trade of the U.S.A. and Canada is carried on through the route.

Importance of Lakes as Source of Power, Irrigation and Drainage: The great lakes of North America are separated by falls and rapids. The Niagara and other falls are used for the generation of hydel power. The fresh water lakes are used for supplying water for agricultural land. The inland rivers, rivers of Asia, namely, Amudaria and Sirdaria discharge their water into the Aral Lake.

Thus we find the importance of lakes on the economic life of man.

Influence of Canals

Canals are excavated by man for various purposes. The canals are mostly used for (i) irrigation, (ii) transportation, (iii) pisiculture and (iv) other purposes. The canals of Punjab, U.P., Rajasthan etc. are extensively used for irrigating the agricultural land. These canals are the life-blood of these states. The canals developed from the river Nile have turned the desert area of Egypt into fertile agricultural land. The canals of France, Germany and Soviet Union play the vital role as inland waterway in these three countries. The canals of multipurpose river projects of India have great contribution in checking flood, soil erosion and developing irrigation system. Other than those mentioned above the artificial lake, natural water bodies, like tank, well, etc., also have great importance on the development of rural economy.

Influence of Climate

Climate as a factor of environment has far-reaching and wide and deep influence on man and his activities. It has direct and indirect influence on man. Climate directly influences man's energy, health, efficiency, food, clothing, shelter, etc. Again by directly influencing vegetative and animal lives, climate indirectly influences man's activity in utilizing these natural resources.

The influence of climate is more effective on the following :

Distribution of population and density of population depends much on climate. The polar region with very cold climate is unsuitable for human settlement whereas the temperate region with moderate climate is highly suited to man's settlement. The equatorial region with heavy rain and high temperature throughout the year is not ideal for human settlement. Hot desert is thinly populated.

Basic needs of man are food, clothing and shelter. Food is mainly obtained from staple crop. Different food crops are grown under different climatic conditions. Wheat is the product of the temperate region whereas rice is principally the crop of the monsoon region. In hot lands fish is the principal protein food but in the temperate climate meat is a principal diet along with cereal food. In the polar region meat and fish are the principal food.

For protecting the body civilized people use various types of garments. In hot climate cotton cloth is common. In cold climate wollen garments are essential. In desert area a particular type of garments are used.

The type of shelter and the materials used for its construction depend much on climate. Bamboo huts and sloping roofs thatched with straw are commonly used in the equatorial region. In polar region ice-built house is used. In developed temperate and tropical areas brick-built houses are common. In the desert region special type of house is constructed.

Economic activities: Economic activities of man are also influenced by climate. In tropical grassland hunting is an important occupation of man. Monsoon climate is highly suited for agriculture. Mediterranean region is noted for fruit-farming. Temperate region is noted for mechanised type of agriculture and industry. The development of industry and transport system in any region is influenced by climate. Climate influences industry directly and indirectly. The raw materials for the industry are obtained from various sources. Raw materials from forest products, agricultural products and animal products are the contribution of climatic condition in any region. Energy and ability of industrial labour is influenced by climate. Production of hydel power depends much on climatic condition.

Transport system in any region is affected by strong winds, heavy rain and flood, snowfall and freezing of river, seas, and accumulation of ice and dense fogs. Heavy monsoon rain and melting of ice in the Himalayan region of India cause flood dislocating road, railway and river transport. Northern and North-Eastern parts of Canada during winter months become ice-bound making them unsuitable for use. Air transport is affected by bad weather condition. Cultural development of any region is also influenced by climate. Architecture has followed the dictates of bright, sunny Mediterranean climate.

Influence of Soil

Soils have great influence on plant life. Without soil, agriculture is not possible. Population is dense where the soil is fertile. Soil forms the basic need of agriculture. Soil gives support to plants and supplies moisture and food to the plants through their roots. Different types of soil contain different types of plant food and for different types of crops different types of soil are necessary. As for example, rice requires clayey alluvial soils, cotton grows well in black sandy loams, tea thrives well in red soils rich in iron and nitrogen. Infertile soil is unsuitable for agriculture. Of course man today increases the fertility of soil by adding manures and fertilizers and dry sandy desert soil is used through irrigation. Soil fertility is also increased by rotation of crop, keeping land fallow for a period after using it for sometime.

Influence of Animals: Animal has great effect on man's occupation. Animal is used as a beast of burden, and for drawing cart and plough. Yak in Tibet, lama in Bolivia, camel in Sahara are extensively used. Animal supplies a great variety of products such as milk, meat, bones, wool, silk, fur, feathers, hides and skins, etc. Animal-hunting. Stock raising, dairy farming, sericulture, etc., are the important occupations of man.

The people of Tundra Region, the Eskiomoes, depend very much on animal—the reindeer. They drink its milk, eat its meat, make tents for living with its skin and its tendrons are used for making ropes. Utensils and instruments are made of its bones.

Of course, today, in the days of mechanisation, use of animals in transportation is gradually declining.

Influence of Vegetation: Vegetation has great influence on man's life. Vegetative life purifies the atmosphere by drawing carbon dioxide and giving out Oxygen.

The Vegetation can broadly be classified into three groups, namely, (i) Forests, (ii) Grasslands and (iii) Shrubs.

(i) Different types of forests are found in different parts of the world. Various types of benefits are derived from forests by man. Even today in some areas forests are the sources of food, shelter and even clothing.

The benefits we derive from the forests are the following:—

- Various types of raw materials such as timber, fruits, rubber, cork, camphor, gum, resin, turpentine, creepers for medicinal purposes, bamboos, etc., are obtained from the forests.
- Some industries depend much on forest products. Soft timber and grasses supply raw materials for paper pulp. Match-boxes, match sticks and packing boxes are produced from soft timber. For making furniture and for other constructional purposes timber is extensively used. Lumbering industry is highly developed near the temperate forest of Canada.
- Forests make the atmosphere humid and thus cause rain. Flood and soil erosion, stormy wind, etc., are checked by forests.
- Forests supply food to man and animal. Fruits are collected for man's use. Roots and leaves are fed by an animals like cattle, sheep and goat.

But in some parts forests stand in the way in developing the region economically. It is very difficult to clear the evergreen forest areas of the ever hot and ever wet equatorial regions of Congo (Zire) and Amazon basins for agricultural purpose. The climate becomes unhealthy for human settlement and development of transport system becomes difficult.

Many areas are occupied by grasslands. Grasslands can broadly be divided into two groups—tropical and temperate. The tropical grasslands are the home of both herbivorous and carnivorous animals. Hunting is the principal occupation there. On the other hand, temperate grasslands of the U.S.A., Canada, Argentina, Australia, etc., are the principal agricultural regions for temperate crops like wheat, barley, etc. Livestock rearing is an important occupation of the people. Fertile soil of temperate grasslands rich in nitrogen is ideal for wheat.

Scrublands and shrublands supply fodder for animals. These grasses prevent soil erosion and approach of desert condition.

Influence of Minerals

Minerals are the backbone of modern industry. The standard of economic development of a country is measured how efficiently it could utilise the mineral resources. Mineral is also the index of civilisation. The gradual change from stone age to atomic age gives an idea about the trend of efficiency and enterprise man could achieve in developing economy. Coal, petroleum, atomic energy supply power to modern industry and transport. Iron, copper, lead, tin, aluminium supply raw materials for metallurgical industries. The highly developed industrial countries of the world like the U.S.A., U. K., U. S. S. R., W. Germany, France, Japan, China are the great consumers of minerals. India's industrial progress after independence is due much to the production and utilisation of her minerals.

Minerals are utilised for the following purposes :

(1) Minerals are used in industry as raw materials and sources of power. Iron and steel, Aluminium, metallurgical industries consume iron-ore, Bauxite, etc., as raw materials and coal and mineral oil as sources of power. Industrial activity of the U. S. A. and U. S. S. R. depends on richness in their minerals.

(2) Modern transport system namely, railways, automo-biles, aeroplanes, ships, etc., depend much on iron, aluminium, tin, etc. Coal and petroleum are used as power. Large quantities of aluminium, copper, etc., are used in electrical industry.

(3) Agricultural machineries are produced with the help of minerals and manures are obtained from potash, phosphates and nitrates.

(4) Minerals are the great sources of foreign exchange. Oil-field of the Middle Eastern countries and iron-ore of India are the great earners of foreign exchange.

(5) Gold and other metals are used for making coins. Minerals are used for making ornaments, jewellery, household utensils, medical and surgical instruments.

But the minerals are not uniformly distributed. So if minerals are not properly obtained economic-activity is retarded. In the political world the minerals have great influence. Arab countries control over oil has great influence on world's trade and politics.

Of the different non-physical factors of environment, Population, and Political and Social Organisations are the most important.

Influence of Population

Population of my country plays the outstanding role in the economic development and acts as the principal resource in shaping the economy of a country. A man with his power of invention, physical and mental energy, culture and quality utilises the resources available under different environmental condition to develop economy. Without his help no activity is possible. A country may be otherwise full of varieties of resources. Unless these are properly utilised by man, no development is possible.

There are many countries which could not be developed properly on account of scarcity of population. The countries like China, India could develop agriculture as a large number of agricultural people are there. The U.S. A., U. K., Japan, etc., were developed in agriculture, industry, trade and commerce by a large umber of people. But a large number is not always the only ideal condition for the economic development. High-level manpower is needed. The advanced countries like the U. S. A., U. K., U. S. S. R., etc., have continuous pressure to innovate and make new discoveries in Science and Technology. The resources are utilized by the introduction of new techniques.

There are many countries in Asia, Africa and South America where the number of people is more than what they can accommodate. But in comparison with the industrially developed countries of the world, the density of population is not high. These countries are over-populated because they could not make the best utilisation of the economic resources. Because of the low level of economic development the standard of living is very low. But by the proper utilisation of economic resources better living facilities can be ensured and economically these countries can be made well developed and the problem of population will be minimized. India, through proper planning and economic development in different sectors can only make the country rich and solve the population problem.

So the number, the quality and culture of the people matter much in the economic development of any region.

Influence of Political and Social Organisations

Political and social organisations of a country have profound influence on the economic life of a country. Though a state has full responsibility to maintain law and order, to ensure security , to promote social welfare through education and care of health of the people yet it has a significant role to play for its economic growth. The socialistic pattern of society which India desires to achieve has become realistic by the adoption of economic planning through public and private sectors. India is heading towards industrialisation and within a few years will be able to solve the food problem which is really an acute problem for the country. A stable popular Government can only shape the economy of a country on planned basis spread over a long period. Efforts on the part of India to improve the conditions of the people through peaceful co-existence with other economically developed countries of the world are really sound and fruitful.

By establishing trade relations and borrowing technical and financial help from the developed countries, an undeveloped country can improve its economy.

Social organisations also contribute much towards the economic prosperity of a country. The joint family system, high birth rate, etc., stand in the way of India's economic progress.

Adaptation of Man to his Environment

Different groups of people living in different parts of the earth differ from one another in respect to their physical and mental characteristics, food, dresses, languages, religion, race, mode of living and above all inherent culture and tradition. Of these, some are, such as body feature, colour, race, religion, etc., hereditary but others are controlled by physical and cultural environments. Some people through their inherent skill and quality, culture and tradition, education and scientific achievement have been able to make their mode of living very high and enterprising and others are even today uncivilised and carry on primitive mode of living surrounded and influenced by the local environmental conditions. As for example, people of the Western

European countries, Soviet Union, United States, Japan, China and India are civilized and advanced group of people. They have developed their agriculture, mining, industry, trade and commerce very greatly. Scientific development and invention have made their life more comfortable and advanced. On the other hand, Pygmies of Central Africa, Bedouins of Arabic and North Africa, the Eskimoes of North America, Samoyed of the U. S. S. R. still live a primitive life depending primarily with local environmental conditions. Though they have come in touch with civilized group of people, yet the influence is very limited and localised.

Though the mode of living of the different groups of people have been profoundly influenced by the factors of environment-physical and non-physical, yet the civilized groups of people have been able to control and modify the factors of environment in their favour. On the other hand underdeveloped and uncivilized groups of people are still the slaves of nature. The way a group of people try to conquer and overcome the difficulties put forward by their environment through their inherent skill, culture and scientific achievement is called the *adaptation of man* to his environment.

Influence of Environment and Adaptation of Different Groups of People

To meet the daily needs, namely, food and drink, shelter, dresses, tools, transport and other amenities of life under dif-ferent enviornmental conditions endowed with different types of culture and quality carry on different types of occupation.How do these occupations vary from region to region with the corresponding change in environmental and cultural conditions are *described below in a nutshell:—*

A. *Primitive occupations,* namely, *Gathering, Hunting* and *fishing:* This type of occupations are undertaken primarily by the primitive people. Their are still some areas where this type of occupation is carried on.

The *Tundra Region of North America* and *Asia, the Northern Mexico, the Congo basin of Africa and Amazon basin* of *South America, South-West Africa* and *Bachuanaland* and in some parts of *Australasia* still have some primitive groups of people who preserve their primitive life.

The Tundra region is occupied by perpetual snow and ice. Agriculture is difficult. The Eskimos and Samoyeds rear Raindeer, live in special type of dwellings made of ice and skin, carry on hunting of animals and catch from the sea seal, whale and other fish. Civilized group of people in North America and Asia are trying to develop the region.

In Congo and Amazon basins ever hot and ever wet climate help the region for the luxuriant growth of evergreen hard-wood forests. Agriculture is difficult. People generally live on tree top and collections of fruits, rubber, and hunting are the principal occupations. Of course advanced group of people have cleared the forest margin for agriculture—specially for growing rubber, rice and sugar-cane.

B. Pastoral Nomadism : The Tropical and Temperate grasslands where the rainfall is very small, agriculture is not possible. Animal rearing and hunting are the most important occupation of the people. The nomads move from place to place with their herd of animals. *Kirghiz* of Central Asia, *Masai* of East-African Savannah, *Kikuju* of Northern Kenya, and *Navaho* of Mexico and Arizona are important primitive groups of people who carry on these types of occupation.

C. Forest Dwellers carrying on *Hunting. Trapping* and Shifting *Cultivation : Equatorial* and *Tropical evergreen forests* of Congo and Amazon basins, Indonesia and forests of South-East mountain areas have hot and wet climate. Forest growth and growth of weeds and creepers very quick. Clearing of forest is difficult. Agriculture is carried on by the primitive people by shifting method. Hunting and trapping are the most important occupation of the people.

In the landforms forests of North America and Eurasia, Agriculture is carried on in limited areas. Lumbering is the most important industry. Far-bearing animals are caught by the people. Far-trading is an important occupation of the people. Trees are cut during winter months and they are either dragged when thawing takes place or through rivers during summer months. Valuable soft timber are highly useful for paper pulp, match boxes, match sticks, furniture etc.

D *Intensive subsistence type of Cultivation of Monsoon lands:* In South East Asia, especially in China, India, Java and Japan rice is cultivated by intensive subsistence type of farming. Ganges Valley & Delta and coastal plain of India, the Irrwadry Valley of Burma, Yangsi-Kiang and Hoang Ho Valleys of China, the river valleys of Thailand and Indo-China are the principal areas of production. Other than rice, wheat, maize, sugarcane, and oilseeds are also grown.

E. *Plantation farming in Tropical Areas :* The white people during the 18th century developed the plantation farming industry with their capital and organisation and local labour in colonies. The principal of them are Rubber plantation of Indonesia, Sugarcane cultivation in

Java and Cuba; Coffee cultivation of Brazil and India and Tea cultivation of India, Sri Lanka and Eastern African countries. The soil and climate are very ideal for the production of these crops in respective areas. Of course the industry is now carried on by local people. These crops are very important earner of foreign exchange.

F: *Mediterranean type of Agriculture:* Summer hot and dry climate and winter wet climate help much in the production of various types of fruits, and wheat, rye, barley, etc. in the mediterranean lands. The principal fruit trees are, citrus fruits (such as oranges, lemons, grapes); deciduous fruits (such as peaches, grapes, the olive, almond and pig).

Olive is the distinctive crop of the mediterranean region. Where there is mediterranean type of climate, there is olive. Production of mulberry trees are also the important product of the region.

The lands around the mediterranean land, the valley of California, Central Chile, South-West and South-East Australia carry on producing mediterranean crops.

G. *Intensive mechanised type of crop :* This type of method is adopted in the densely populated regions of Western Europe and Soviet Union and in the temperate grasslands region of the world for the cultivation of wheat, sugar-beet, etc. Nile delta produces rice by intensive method of cultivation.

H. *Animal Rearing:* The tropical and temperate grasslands are used for rearing cattle and sheep both for meat, hides and skin, wool and milk.

Man's Impact on Changing Environment

The vital role of man in ecology has been stressed within several topics considered so far. This last chapter summarizes man's past impact on British ecosystems and presents two detailed regional examples. The ecological activities of man are a containing phenomenon and present and probable future patterns of these are also examined in ecosystem terms. The difficulties arising from multiple land use and mounting recreational pressures on our countryside are then considered, particularly with respect to the conservation problems they pose for environmental planners.

The Legacy of Human Interference

From about 5000 B.C. onwards man has had an ever-increasing influence in determining the distribution and many of the properties of our flora, fauna and soils. The early cultures, Palaeolithic, existing as hunters and gatherers at low population densities, were once thought to have produced only slight changes in their environment. However, for certain unpland habitats, such as the North York Moors and Dartmoor, Simmons has argued for a clearly detectable. Mesolithic impact and there is increasing evidence for this view. There was also impact on those marginal lowland sites which were ecologically susceptible to change by virture of their pedological characteristics.

By about 3000 B.C. Neolithic man had arrived in this country and his influence was certainly pronounced. As a cultivator, he cleared mole forest and to protect his own grazing animals and increase his food supplied he began to reduce the competing native fauna of the woodlands. We have strong evidence that selective clearance of some tree species was taking place at this time (young lime and elm trees to supplement scarce grass supplies for domesticated cattle) and some slopes were experiencing soil erosion as a result of use by man. The first Neolithic forest clearances were temporary plots which were subsequently abandoned reverting by subseres towards the climax forest composition. Before long, however, more permanent clearances were established and by then much of the forest in some regions would have been essentially secondary in form. This was particularly so on the better drained soils and slopes. It was initially thought that a high concentration of prehistoric remains on limestone scarplands (*e.g.*, the Chilterns and North Downs) and uplands (*e.g.* Dartmoor, Pennines and Forth York Moors) meant that early man avoided lowland sites. But recently many more prehistoric settlement sites have come to light in the lowlands causing a revision of this view. During the following Bronze Age and Iron Age the forest underwent further reduction–Bronze Age settlements, for example, reached quite high levels on Dartmoor.

It was these cultures that the Romans encountered during their invasion of Britian. In this conquest they set fire to woodland as part of their military strategem against local populations. The Romans were also important in clearing lowland areas for cultivation. They were able to tackle some sites which would have proved unattractive to earlier cultures whose technology for dealing with the denser vegetation and poor drainage of the wettest valley bottoms was limited. Throughout this and earlier periods the British flora was erceiving additional species and Godwin lists many introductions which are now largely accepted as native. These include Castanea sativa (sweet chestnut), Papaver rhoeas (poppy), Juglans regia (walnut) and Sinapis arvenis (wild mustard).

Forest reduction must have continued throughout the Roman period and into the Dark Ages but there would still have been vast tracts of woodland landscape. The next clear evidence of profound environmental alteration comes with the arrival of the Saxons. From about the sixth century onwards, these agricultural colonizers opened up much of the lowland forest and established perhaps 90 percent of the existing pattern of English villages and hamlets. The process used was 'assarting' or grubbing up and burning trees and shrubs to prepare the land for tillage. About the same time, Norse, Danes, and Viking first raided and later settled many parts of nothern Britain. They, too, used fire extensively to establish their rule. Under the Norman feudal system, the clearance and reclamation of land continued, but large areas were also set aside as royal preserves—open 'wastes' and woods suitable for game hunting, such as the New Forest. However, their introduction of the rabbit and their fostering of deer herds did much to restrict the regeneration of trees.

These destructive trends gathered pace during the Medieval period and locally timber supplies became very scares. The Exchequer Rolls of Scotland show Baltic timber being imported to Edinburgh as early as A.D. 1329. By the seventeen century, Samuel Pepys, John Evelyn and other prominent writers were bewailing the dearth of timber in Britain. Much in the south had been used for fuel, buildings, ship construction and charcoal production. Increasingly, merchants moved into remote regions for their supplies (*e.g.*, the Lake District, the Loch Lomond Woods). On Roman maps the simple descriptive term Coledonia Silva—the Caledonian Forest-had been boldly written across Scotland, yet in A.D. 1617 Sir Anthony Weldon, touring lowland Scotland, was led to the comment, 'there's scarce a tree to hang a Judas on, and indication of the completeness of tree removal in these parts.

Regional Examples

The human impact on the natural environment can be followed more specifically if we take two detailed examples. The first concerns the pine and birch woods flanking the Cairngorm Mountains of Deeside and Speyside. They are remnants of the ancient Caledonian Forest. In complete contrast are the numerous small mixed-deciduous woodland plots lying a few km east of Leicester, many of which are fox coverts. Some of these are much altered in character but originally were part of the once extensive Summer Deciduous Forest. Others, however, were entirely created by man as planted woodland for fox-hunting. Where are most present coniferous woodland in Britain in quite obviously planted, so much of the deciduous woodland is equally ' artificial'.

The Cairngorm Forests

These forests formerly extended to altitudes of 620 m or more but today the upper limit seldom exceeds 490 m. Much of the forest is now open ground or forestry plantation. The gross disturbance of these woods began about A.D. 400 in Pictish times when large tracts of lower forest were cleared for primitive cultivation. The decline of forest on the higher slopes took place much later, mainly in the Medieval period, with the evolution of cattle-breeding, the main-stay of the Highland economy until the late eighteenth century. Gaffney has shown that by the sixteenth century most forested land between 300 m and 600 m had been turned into highly prized grazings, Through the area ran several important drove roads for moving the cattle to summer pastures (shealings) or to lowland markets. These routes attracted rivers or thieves who were extremely active until the military set up control points in the region. In addition to the local cattle on the summer pastures, there were ' strangers' or gall cattle. For local tenants it was common practice to take in such cattle for a fee from farms as remote as 35 km away. We cannot be certain just how well wooded the higher ground originally was but there are numerous references in the Gordan Castle Estate Papers to unlawful grazing and heavy exploitation timber at these levels. It is clear, however, that the trend was always towards continual reduction of the forest cover.

Other animals, in addition to cattle, featured in the economy and ecology of the pre-sheep period of this part of Highland Scotland. Deer were numerous and greatly encouraged by the Dukes of Gordon, the upper glens eventually becoming over-stocked. In winter their traditional feeding grounds were the native pinewoods. Horses, goats and sheep were also numerous. Macfarlane in 1748 described the upper Dee as richlystocked with these animals. In 1779 the Forest of Glenavon was 'eaten up like a sheep pasture' and in 1794 'the forest contained many green spots and...affords pasture for a thousand head of cattle'. It is obvious that these animals would have a marked effect on tree seedling regeneration and ground flora composition.

Several activities associated with called husbandry also had a pronounced influence on the woodlands. First, the construction of temporary bothies ('scalans') at the highest pastures required the use of 'divots and trees'. Secondly, he law, cattle had to be confined or 'hained' at night and this often necessitated the building of timber enclosures. Thirdly, crops needed protection from grazing animals and this was often achieved by erecting timber barriers ('gathering'): at one site 5,000 young trees were used to protect a cornyard. A measure of this destruction of native trees is given by the petition of James Grant of Grant to the Parliament of James VI 'that fir wood in Abernethic and Glencherneck are dayly cutt, stollen and carried away by the tenants of Strathspey... and that the birk-woods are wolly destroyed by peiling of the timber peiled standing rotting in the woods... and against the great hurt and prejudice done to the fir woods of Strathspey by cutting and destroying standing trees for to be candle fir to all the inhabitants'. (During times of hardship peeled birch bark was used as a human food source. The 'fir' mentioned is Scots pine, candle fir being a primitive form of home lighting.)

The practice of muirburning was of even greater ecocogical importance. This was widespread and frequently the fires got of control. It because so excessive that by 1695 severe penalties were enforced for burning too near the pinewoods—a third offence merited hanging. By the end of the eighteenth century large-scale sheep farming had replaced cattle farming but grazing and muirburning continued as dominant activities working against the remaining forest over. In the Cairngorms sheep were never so important as in other parts of the Highlands and eventually most sheep runs were converted to grouse moors. Again, this meant the continuance of regular muirburning.

The eighteenth century also saw the intensive commercial exploitation of the remaining woods by the timber companies, The York Building Company of Hull was responsible for the main onslaught in Abernethy Forest from 1728 onwards. Similary, Glenmore Forest was worked over in 1783 and the adjacent Rothiemurchus Forest at this time was proving $10,000-$20,000 annual timber revenue. Several severe forest fires occurred in these woodlands during the eighteenth and nineteenth centuries and there was further jheavy timber exploitation in the World Wars of the twentieth century.

The nobility have acted unwittingly to preserve some parts of these forests through their strict ownership and sporting interests. In the nineteenth century there was extensive tree planting by the nobility and one Gairngorm parish could boast 14 million 'firs' planted by 1877. However, the main plantings seen today are due to the Forestry Commission who, from 1923 onwards, began a comprehensive re-afforestation programme using several exotic conifers (mainly spruce, lodgepole pine, larch and fir). Glenmore became a National Forest Park in 1948 and in 1954 the large Cairngorms National Nature Reserve was declared.

Many of these recent measures to repair the ravages of previous centuries are now placed in jeopardy by the current massive influx of tourists, a phenomenon which we shall examine shortly. Recreational and sporting activities, forestry, conservation, and grazing, all operating together at these altitudes must lead to a severe clash of interests which will have to be resolved successfully if these already much modified woodlands are to survive in an acceptable form.

The Woodlands of East Leicestershire

The subdued, rolling landscape of East Leicestershire (which now included the former country of Rutland) stands in sharp contrast to the grandeur of the Cairngorm Mountains. The East Midlands are an important agricultural region but, nevertheless, the area contains a surprising number of small scattered woodland plots: 130 are shown on the O.S. 1:25,000 Leicester (East) SK 60/70 sheet. Many of these are less than 200 years old. Nearly all owe their form and existence to the fox-hunting activities for which the country is famous: the natural Summer Deciduous Forest which clothed the region in prehistoric times has long since given way to farming.

In the twelfth century all of Leicestershire and neighbouring Rutland was declared 'forest'. *i.e.*, wooded and open waterland reserved for hunting. Leicester became exempt in A.D. 1235 but Rutland did not until much later . Some of the present-day woodlands around the parish of withcote, although now much altered by man, may be directly descended from the extensive early Medieval Leighfield Forest that was retained in this part of the Midlands until relatively late. The largest of them, Owston Wood, once grew very fine native oaks but these were clear-felled after the First World War. This wood was neglected for 35 years, acting simply as a source of stakes and fence material for the Cottensmore Hunt and local farmers. In 1960 the Forestry Commission took it over but have had continuing problems with weed species (willow herb, bracken, bramble and, particularly, willows) that became well established during the period of neglect. Another difficulty is that the heavy Lias Clay soils puddle badly in winter and crack deeply in summer, this condition being made much worse by the constant churning of the Cottesmore Hunt who draw these woods each winter.

Billesdon Coplow is even more of an artefact since much of the hill consists of nature planted beech reaching 30 m or more. The wood also has a larch plantation and open areas are planted but with a mixture of Scots pine and beech. A few of the original oaks remain. Botany Bay Fox Covert, adjacent to Billesdon Coplow, was originally planted soon after the time of the Australian settlement, as the name suggests. It is mainly degenerate oak-ash wood with an open canopy which has encourged profuse growth of hawthorn, hazel, blackthorn, privet and elder shrubs. The hazel was formerly coppiced but some shrubs are now 'chopped' to promote a tangle under growth suitable for fox cover.

Knighton Spinney, now surrounded entirely by suburbs lies 4.5 km from the centre of Leicester. It is an entirely man-made woodland and many other examples like it could be quoted from the East Midlands. Formerly the site was ploughed land but in the early nineteenth century it was planted with rows of oak. Ash, elm, sycamore and willow are now also present. The wood is rather neglected and has a poorly developed ground flora. It is no longer drawn by the Fernie Hunt and now forms the core of a municipal park in the Leicester suburbs.

In the late nineteenth century, at the height of fox-hunting as an acceptable and overt social phenomenon, over 300 horses might leave centres such as Market Harborough or Melten Mowbray for a day's sport in these woodlands no. Today, despite antagonism from some sections of society (a topic discussed later in this chapter), fox-hunting has not diminished and there are more hunts than at the turn of the century. In a landscape which is predominantly agricultural, it will be interesting to see what happens to these small but important woodlands in the future if rapidly rising costs and the present pace of social change leads to a decline in fox-hunting. Already many are under increasing pressure from tourists, and in several, vandalism is evident. Like the remote Cairngorms, these woodlands represent a dynamic situation. In both cases there is overwhelming evidence of man-induced changes during the past and a certainty for the future that man will become more not less important in the ecology of these areas. These two regional studies could be matched by similar examples drawn from virtually any part of the British Isles. As we shall see in a later section, for all British woodlands the basic problem is to so manage this resource that the pressures emanating from the various human uses do not undermine good conservation practices.

Much that we cherish in the countryside today is artificial and results from centuries of environmental modification by man. The open hillsides of the Lake District, the Chalk Down-lands with their distinctive chalk floras, the expanses of neither moor and the 'typical' rolling English countryside of fields, hedges and small woodlands all fall into this category. Some species have proved resilient to change and others have adapted well, extending their range into man-made environments. The seagull, for example, appears to be just as much at home scavenging from inland urban refuse tips and frequenting reservoirs as it does in its more usual costal habitat. Likewise, pigeons, closely related to rock doves, find suitable nesting sites on the 'cliff ledges' of office buildings. An even more interesting case is the collared dove (Streptopelia decaocta). This native of India reached Turkey in the sixteenth century but has now shown an explosive spread across Europe during this century. First reported in Lincolnshire in 1952 it had reached a population of 30-50,000 in Britain by 1970. Coombs *et al.* report that it is invariably associated with human settlement, particularly favouring town parks and suburban gardens. The introduced Oxford ragwort (Senecio squalidus) is an example from the plant world. It has spread rapidly in Britain along the cinder-strewn embankments of railway and into other disturbed stony sites where conditions approximate to the lava fields around Mount Etna, one of its original habitats in Sicily. An increasing number of species, however, are highly vulnerable to these changed landscapes and some have become extinct while many are now rare. These trends must be carefully considered as we contemplate present and future patterns of activity in rural Britain.

Present and Future Impact

The increasing pressure of recreational pursuits on the countryside was previously mentioned and this will now be examined in more detail. The developments described mainly relate to the Cairngorm Mountains but most of the findings are of general application because many other parts of the British Isles are undergoing a similar experience, *e.g.* the Lake District Dartmoor and the Peak District.

In the Cairngorms the hill-walker and mountain-climber, always keenly appreciative of what the region offers, have now been joined by rapidly growing numbers of motorised tourists. For some time tourists had limited access to the region but now there have been road improvements and a great extension of tourist facilities. The small settlement of Glenmore has been expanded and the road widened and continued to 760 m. Chair-lifts, ski-tows, snack bars and restaurants now exist at the terminus. These early investments soon bore fruit. On Easter Monday 1962, 5,000 people were estimated to be on Cairngorm itself and during the first seven months of operation 80,000 people used the chair-lift. Over a five-day period in 1965 the Cairngorm Board collected some $4,500 on lift and toll charges. Commercial success like this prompts more investment, an example being the nearby $2 2/1 million Aviemore Centre which includes two large hotels (now frequently used as conference centres), chalets, a swimming pool, an ice-rink and artificial practice ski-slopes. In 1982 further proposals for expansion into adjacent corries were under consideration (including two ski-lifts, two tows, parking for 1,000 cars, cafeteria, toilets and snow fences). Lickorish, writing about planning for recreation and leisure, noted the point made by the US Department of Commerce as early as 1950: when a community can attract a couple dozen tourists daily throughout the year, it is economicaly comparable with acquiring a new manufacturing industry whose annual payroll is $100,000. And business interests soon realise that an investment of £100,000 in a mountain that brings in £10,000 profit a year is just as useful as the same money made from more conventional investments. However, some might see the situation rather differently in that such schemes may not create as many jobs as the same money more conventionally invested and therefore the end result might be less socially acceptable.

An increasing number of visitors to the Cairngorms are people on day trips. Many will come in their own vehicle but there is a rapidly growing bus and coach tour element in these numbers (this expansion in the early years of these recreational developments can be seen from Scotland's bus and coach tour passenger number which increased from 3,900,000 in 1960 to 5,800,000 in 1963). Because of the relative remoteness of many recreational areas this implies a lot of travelling which leads to congestion on roads originally built to serve small, isolated Highland communities.

In the context of the United Kingdom, Dower, writing on the function of open countryside, has considered some of the main conclusions of the 28- volume report published in 1962 by the Outdoor Recreation Resources Review Commission of North America. The main American findings were:

1. The population of the United States will double by AD 2000 but the national demand for outdoor recreation will treble.
2. The recreational areas are remote from the centres of population.
3. Driving for pleasure is today the most popular outdoor activity.
4. Most people try, to centre outdoor recreational activities around a body of water.

It is more than 20 years since Dower's paper and no such detailed comprehensive report on recreation exists for this country. However, an increasing number of specific areas within the British Isles have since been produced (*e.g.* Moss; O' Riordan) and several general texts, often viewing the recreational problem in global terms, contain much that is relevant to this country (*e.g.* Simmons; O' Riordan and Turner). It is now quite clear that the pattern of leisure-time activities in Britain has moved strongly in a similar direction to that seen in the United States. We already know that the majority of motorised tourists seldom venture more than a few hundred yards from their car and this often gives rise to the 'honey-pot phenomenon' where large numbers congregate at popular vantage points. A good example in the Cairngorms is Loch Morlich, where erosion of the shoreline is already a problem (similar shoreline erosion has been reported by Tivy and Rees for many Scottish lochs). There is a strong case for preserving extensive parts of the Cairngorms against massive penetration by tourists, but although a large Nature Reserve does exist the pleasure-seeking public have never shown great respect for lines on maps. Such a vast influx of tourists coupled with unplanned piecemeal development will do much to destroy the essential character of these mountains.

Problems of Multiple Land Use

There is agreement that development in regions such as the Cairngorms must be planned and should consider all interests, including the larger national interest. The problem is to find a plan broadly acceptable to all and, at the same time, capable of implementation at reasonable cost. Difficulties arise because interests clash and priorities have to be determined. One interest, that of the tourist, has been mentioned but others include those of the water boards, the power stations, the military, mining, forestry, agriculture, sporting activities, transportation and last, but not least, conservation. Many of these uses are mutually exclusive and nearly all have their problems of operation greatly increased by the presence of large numbers of tourists.

Another factor adding to the difficulties of resolving clashes of interest in our countryside is the question of land ownership. Much of our landscape is still privately owned and the land-use pattern here will often, but not always, be determined by the necessity to make a profitable return on investment. Large areas are state-owned or state-controlled and motives other than profitable return may then decide land use. But, it should be remembered, many designated National Parks do include farms and private land within their boundaries. This makes a comprehensive development plan for the whole region that much harder to achieve.

It is beyond the scope of this book to deal fully with each interest in turn. Instead, a few comments under each heading will be made so as to convey something of the complexity involved in multiple land-use planning. However, fuller treatment will be given to conservation because of its central role in preserving as a resource these very landscapes that we wish to use and because effective conservation calls into consideration many of the ecological principles outlined in previous chapters. This makes it of special interest to geographers.

Water Boards

The industrial and domestic demand for water has risen sharply in this century and it is usually argued that this will continue to be the case in the foreseeable future. For example, the installation of washing machines and dishwashers in the home and the rise in personal cleanliness have led to a dramatic increase in household consumption of water. Agriculture now uses more water and consumption is likely to expand even further in eastern England, where intensively grown crops are likely to benefit from the use of sprinklers and irrigation three years out of every five. Heavy demand, coupled with a very variable climate, already leads to local and regional shortages during spells of dry weather. However, lorecasting future demand for water is not without its difficulties. Several large traditional industries which are heavy users of water are now in marked decline (*e.g.* steel) Many of the newer industries based on electronic technologies will not require massive inputs of water. The problem is well illustrated by the Kielder Reservoir. Since completion of this recent £167 million scheme the industrial demand in the Northumbrian Water Authority area has already decline by 5 percent and full use of the water from Kielder is not now likely until the year 2000.

The creation of large reserviours must take place at topographically suitable sites which, by their very nature and location, will tend to be attractive areas for recreation. Many reservoirs have been sited in our uplands but an increasing number are now in lowland areas. Wherever the site is, local opposition to the loss of land involved can be passionate: a group opposed to a new reservoir on farm land in the Roadford Valley, west Devon, even threatened deliberate pollution of the River Tamar if the scheme went ahead. Extensive areas of farm or forestry land have been flooded by schemes and in the past the feat of water pollution and bank erosion has usually meant the exclusion of people from these sites. Fortunately, this policy has been revised of late and more reservoirs now cater jointly for water and recreational needs. This is well seen at the new Rutland Water Reservoir which lies in the heart of an important agricultural region. The loss of farmland here will be partly compensated for by the provision of recreational facilities, particularly for fishing, boating and water-sport interests but also providing picnic areas and nature reserves. However, at Cow Green, Upper Teesdale, plans to build a reservoir to serve the ICI industrial complex on Lower Teeside met great opposition. This was because the proposed site was the habitat of several rare plants, a unique arcticalpine floral element containing species such as Gentiana verna and Minuartia stricta which are confined in Britain to the Upper Teesdale area. Despite a well organized national campaign and strong support from many internationally eminent botanists the plans were implemented.

Another activity by water boards that threatens wildlife is the drainage of wetlands. Halvergate Marshes in Norfolk is a classic case where controversy between farming and conservation interests has raged for several years now. Drainage would allow profitable grain crops to be grown, but to the conservationist Halvergate is one of the last pieces of species-rice grazing marshland in Britain.

Military Training

The Ministry of Defence controls several large regions in the British Isles where the general public are excluded. Unfortunately, some of these regions are highly attractive landscapes with great recreational potential. Examples are western Dartmoor, parts of the Dorset coast and a portion of Breckland in Norfolk. Currently, these areas are under review with strong pressure for more to be released for enjoyment by the public. Some people have argued for an almost total reduction of military involvement in these regions. However, if we spend thousands of millions of pounds on defence it then seems illogical to deny proper training facilities. The importance of adequate preparation in the use of modern warfare technologies has been emphasized by the recent Falklands conflict.

In some cases the presence of the Military actually assists conservation. For example, access to most of the Pendine Sands of Carmarthenshire is prohibited because of a missile and gunnery testing range. Yet these dunes may well act as a refuge for native fauna and they do carry a fuller and more varied vegetation cover than adjacent sections which are regularly and heavily trampled by tourists. Likewise, the Chemical Defence Establishment at Porton Down in Wiltshire controls about 2,000 ha of chalk grassland which is now one of our most important wildlife conservation areas. It is a diverse region which acts as a sanctuary for many interesting species that are rare or absent in the surrounding zone of intensive agriculture.

Power Stations

By the very nature of their activities nuclear power stations must be sited at some distance from major settlements. The same applies to several of the associated Atomic Energy Authority Establishment (*e.g.* Dounreay, Windscale (now Sellafied) and Calder Hall). Because of this, many occupy areas that would normally be of prime conservational or recreational interest. This is particularly so since the vast majority of nuclear power stations also occupy coastal sites (*e.g.* the Central Electricity Generating Board's nuclear power stations at Dungeness, Bradwell, Sizewell, Hartlepool, Hunterston, Heysham, Wylfa and Hinckley Point). Not only are they massive structures in themselves but associated facilities and security considerations mean that a large area of coast is effectively prevented from being used for any other major purpose. Several of the possible sites for future pressurized water reactors (PWR's) of the Three Mile Island type are also coastal locations (*e.g.* Stakeness on the Moray Firth and Gwithian near Land's End).

Mining

Mining and quarrying are dirty, dangerous and unsightly operations and their intrusion into areas of natural beauty is to be deplored. They also generate a form of transport (the heavy truck) which is particularly damaging to the countryside and infuriating to other road users. Nevertheless, we need these raw materials for the very things we may value in other environments, such as good roads, important urban buildings, gravel driveways and garden rockeries. In the long term their worst effects on the landscape can be alleviated by insistence on proper soil restoration measures and the use of the flooded surface pits for recreational and conservational purposes. When carried out successfully, this can actually enhance the landscape. A good example is seen in the Cotswolds where some of the abandoned and now flooded, shallow, surface quarries are used as centres for fishing and boating, while others are reserved for wildlife. The complex will eventually cater for many needs in an area formerly lacking in surface water facilities. Another example of cooperation is seen at Thrislington, Co. Durham. Here, the Nature Conservancy Council and the Durham Conservation Trust are working in conjunction with a quarry company to move a 8 ha field complete with topsoil, rare plants (including several orchids) and animal life to a new site half a mile away. This will allow pure dolomite, an important ingredient in the steel industry, to continue to be mined.

Mining and quarrying are not always confined to hard rock areas, most of which occur as our uplands. Extensive lowland tracts have been worked for gravels, sands and brick clays, leaving numerous scars on the landscape. Only recently has the full recreational potential of some of these sites been appreciated. However, the picture is complicated because some lowland areas fall within National Parks. A case in point is the highly attractive Mawddach Estuary, which is near sea-level yet lies within the Snowdonia National Park. Because of this, plans to dredge for alluvial gold in the estuary aroused much protest.

Forestry

Many British woodlands are privately owned and contribute a very desirable element to our landscape. Whether they remain in their present form may well depend on government taxation policy as much as on anything else. In the past, governments have encouraged tree planting on private estates by a series of financial incentives. But any changes in capital transfer taxation following death of the owner could lead to a sharp decline in many of these woodlands. As table below shows, the private sector still produces nearly 60 percent of our cut timber and is responsible for the vast majority of our more attractive hard wood species.

Table: *Volume of Timber Cut Annually in Britain: Values in Millions of Cubic Metres (Modified from Rooke, 1974)*

Source	*Softwoods*	*Hardwoods*	*Total*
Forestry Commission	1.55	0.04	1.59
Private sector	0.82	1.30	2.12
Total	2.37	1.34	3.17

The state owns vast areas of forest, mostly as coniferous plantations. Formed in 1919, the Forestry Commission was charged with the task of building up strategic reserves of timber following the depletion during the First World War. This has been a difficult task because of the long time-span for tree species to reach maturity and the poor growing conditions at many upland sites. The Second World War also caused further setbacks. However, the Forestry Commission is on target for its aim of 5 million acres (2 million ha) of afforested land (including both state-owned and privately owned). Nevertheless, we still import vast amounts of timber despite increasing home production. Wood and wood products are Britain's third largest import, costing about $ 2,370 million in 1978. Russia is a main supplier which could make our position vulnerable at times of international crisis.

Net imports are unlikely to drop much below 90 percent of our total timber consumption during this century.

***Table:** Consumption, Production and Import of Timber in Britain at the Present and Expected for the Near Future: Values in Millions of Cubic Metres (From Rooke 1974).*

Year	*Expected Consumption*	*Home production*	*Expected net imports*	*Net imports as % of total*
1970	45	3.5	41.5	92
1975	51	4.5	46.5	91
1980	57	5.0	52.0	91
2000	82	8.7	73.0	89

One of the major difficulties with forestry as a land use is forecasting future trends within the industry because of the long-term nature of the crop. Raup has studied the basic concepts and assumptions underlying North American forest management and found them to be false. It looks as though a better assumption would be that predictions beyond one or two decades were more likely to be wrong than right' and 'uncertainties in the long run call for the greatest possible flexibility in resource use'. These lessons may have strong implications for British forestry. One of the major problems in Britain is the vast acreage now planted with fast-growing conifers, mainly Sitka spruce (Picea sitchensis). Sitka grown under our climate is largely a low-grade softwood timber and there is increasing difficulty in finding suitable and profitable outlets for it. While the need for more forest cover can be readily demonstrated, especially when given forecasts of a massive rise in world timber consumption and severe demands. being made on diminishing natural forests (see Bowman), technologies may change and so too may fashions for certain timbers. Thus, what we plant now may not be what is required some 50 or more years later. However, it is the tourist who is likely to cause the main changes in future forestry developments.

In early Forestry Commission plantations the public were discouraged, partly because of fire hazards and partly by the monotonous nature of the vegetation itself. But the key to present changes is contained in a declaration from the Seventh. World Forestry Congress (1972); 'Foresters recognise that forestry is concerned not with trees but with how trees can serve people...His allegiance is not to the resource but to the rational management of that resource in the long-term interests. of the community'. Forestry makes its greatest impact on the general public through the landscapes it creates. The Commission now realize that social and environmental demands must be harmonized with the need to produce timber. By 1974, the Commission had designated seven Forest Parks with a combined area of 176,600 ha and was investing over £1 million. annually in recreational provisions. The Commission had about 14 million day visits to its forests in 1980. In Holland, 75 percent of all state forests now cater for recreating and Britain is also moving rapidly in this direction.

Agriculture

Farming has often been in competition with forestry, particularly in upland areas where clashes also exist with some of the other forms of land usc discussed. Indeed, as we have seen, most agricultural land has resulted from the clearance of forest at some time in the past. In 1972 there were some 11,000 hill and uplands farms in England and Wales. Much upland terrain is marginal for agriculture and may provide a better economic return and more rural employment under forestry. But the position is not simple, as James shows in his review; each case has to be carefully considered and conclusions applied to whole areas or even whole farms may be incorrect. Decision-makers have fallen into the trap of regarding agriculture in our uplands as a homogeneous activity when, in reality, some parts of an area or individual farm are more suited or less suited to agriculture than others.

We no longer live in a world of cheap food supplies (despite EEC surpluses) and a case can be made of either extending farming acreage or greatly intensifying productivity on land we already use. In lowland areas at least, intensive agriculture will increasingly involve large capital investments for complex technological system in farming–an aspect which at best makes the presence of too many tourists a nuisance and at worst a positive danger, according to Bonham-Carter. In some areas, however, the farming community has adopted a more welcoming attitude to the tourist. Along certain stretches of coast and in some unplands, farmers were not slow to realize that a field of caravans can be more profitable than one full of cows. On the flanks of Dartmoor a very large component in the income of some farms is now derived from activities associated with the tourist industry.

Government policy (*e.g.* guaranteed prices or the addition or removal of subsidies) and EEC desisions may well tip the economic balance of marginal agricultural areas one way or another, sometimes strengthening sectional interests and sometimes weakening them.

Schemes such as the drainage of wetlands may benefit the farmer, but seldom the taxpayer. Increased production is often in commodities already in surplus (*e.g.* milk, grains). This means that through the EEC's Common Agricultural Policy these surpluses are brought into intervention to maintain the price paid to the farmer (*i.e.*storage, until sold later and usually at a loss to the taxpayer). This largely explains the current crop of land use conflicts between farmers and others (mainly conservationist), *e.g.* Baddesley Common (Hampshire), West Sedgemoor (Somerset), Romney Marsh (Kent), Berwyn Mountains (Wales). The National Farmers' Union have responded by pointing out that an area the size of the Isle of Wight has been lost to British farming in the last 20 years and that 650,000 people are directly employed in agriculture and some 2 million (8 percent of the labour force) rely for their jobs on British farms and their produce.

Sporting Activities

In the countryside the fostering of sporting interests has played a large part in shaping the fine detail of the landscape. In addition to the numerous fox coverts in areas like Leicestershire, many plant communities were established in both upland and lowland districts specifically for game purposes, *e.g.* grouse moors, pheasant and partidge coverts. In some remote areas the revenue from sporting activities plays a significant part in the local rural economy. A recent survey of countryside sports shows that 4 million people go fishing, nearly 600,000 go shooting or stalking and 214,000 follow hounds, either on foot or in the saddle, at some time during the year—and these people come from a complete cross-section of the British public. These activities account for about £1,000 million of direct annual expenditure and employ either directly or indirectly about 90,000 people. However, not all look favourably on these activities. The 'blood' sports (involving hounds) are now under great threat. The Co-operative Wholesale Society, the largest independent farmland owner in Britan, has now banned fox hunting over its 12,000 ha and so have certain Labour-controlled councils. Such a ban was also included in the recent Labour Party election manifesto despite claims that hunting actually ensures the preservation of many woodland plots and the hunted species. Thus, as in most areas of land use conflict, a political factor must be recognized.

Catering for certain types of sporting interest is largely in private hands and takes place on private land. Good examples are grouse shooting and salmon fishing. However, there is an increasing development of sporting facilities of the general public and this is having its on our countryside. Today, large sums are spent on sporting equipment and there has been a great diversification in sporting pursuits. Although some activities, such as water-skiing or hang-gliding, appeal only to a minority, the clashes with other countryside users which will result from their expansion will soon have to be resolved. The creation of special countryside areas just for active sport is a legitimate use since these activities may do much to relieve the tensions and aggressiveness generated by sedentary or monotonous urban employment.

Transportation

For the foreseeable future we are mainly thinking in terms of road networks when transportation in the countryside is considered. In rural areas which attract many tourists the road network soon becomes heavily congested, having been established initially to meet only local needs. Any major improvement in the network may only result in more people flooding into the region. The better access the available may cause a greater strain on many of the other facilities in the district. So, to a certain extent, these improvements' can be self-defeating. For example, motorways recently constructed in southern England now give quick access from the populous south-east to the popular south-west and there is a danger of 'killing the goose that lays the golden egg'. A similar fear has been expressed for the Lake District in Cullinghworth's review of these problems.

The construction of major road systems means a loss of much land already in other uses. Gradient considerations dictate that this will often be the better quality agricultural land. By their very nature these major roads will usually follow the most open and obvious routes into a region and will be visible from many vantage points. While improving access from outside, they may cause local resentment and a breakdown of established patterns of social association for those living permanently within the region. The controversy over the proposed Aire Valley Trunk Road in Yorkshire stems from such local resentment. Here, a major planning enquiry was repeatedly disrupted by local pressure groups and finally had to be abandoned after costs of about £30,000 had been incurred.

Improvement of minor roads and trackways provides further access for tourists but may also bring benefits to agriculture. As Jones points out, one of the most serious physical handicaps to upland pasture improvement is the lack of good access roads. Millions of hectares are not fully developed because vehicles carrying lime and other fertilizers just cannot reach these sites.

Another major threat to rural areas is the construction of large airports. With their voracious appetite for land, their very specific site requirements and their attraction of numerous ancillary facilities and activities, they represent a major source of land use conflict. One has only to think of Maplin Sands and Stansted to be immediately reminded of these conflicts.

Having seen something of the problems that face the environmental planner we may now turn to a consideration of conservation, bearing in mind that each activity above causes special difficulties for the conservationist. When several are combined they may make good conservation practice virtually impossible: a neat exemplification of this is contained in the two well-illustrated papers recently produced under the general title 'Alarm call for the Broads', (Moss; O'Riordan).

3

The Cultural Landscape

Emergence of Man and Races of Mankind

Among the most fundamental differences which distinguish societies from one another are those which are connected with man's physical nature: the basic division of mankind is into races. It will be argued later that the real significance even of race is social, but for the moment it is necessary to begin our classification of mankind along these purely anthropological lines.

What is race? The word is much maligned. It has been so wrongly used in recent history that it is difficult to divest it of distasteful associations. But it must be retrieved so that it can be used in its proper scientific sense. A constant difficult is the lay habit of using the word loosely and lazily for so many different concepts. People will talk of the human race, meaning mankind as a whole: of the Jewish race when they mean the Jewish people of religion: of Latin races when they mean people with a common group of languages.

A race is a major division of mankind and is defined solely by physical characteristics. It is the final subdivision in the animal world of a series of biological classes beginning with the orders Primates. There are three suborders in the Primates, one of which is the Anthropoidea. This suborder is further divided into two divisions; one of these gave rise to the South American monkeys, the other gave rise in Africa and Asia to several families. One of these, Simeidi, included the Gibbon, Orang, Gorilla and Chimpanzee; another, Hominidae, as the name suggests, was human-like. This family had a unique combination of five characteristics which ensured the evolution of man:

(i) erect posture;
(ii) free-moving arms and hands;
(iii) sharp focussing of the eyes;
(iv) a superior brain; and
(v) powers of speech.

Many kinds of man emerged from this family, some of them possibly over-specialised in one way or another and subsequently dying out. These are known for their fossil remains, and among them are Java man (Pitchecanthropus, erectus), Peking man (Sinanthropus pekinensis), and South African ape-man (Australopithecus). Some of these were similar enough to modern man to belong to the same genus, Homo; such was Neanderthal man (Homo neanderthaliensis), Rhodesian man (Homo Rhodesiensis), and many others. It is possible that it was the least specialised of all which eventually gave rise to the species Homo sapiens, and recent discoveries in Kenya suggest a very long history indeed to this species.

This long lineage emphasises that man is an animal and belongs to a classified system which is common to all animals. Homo sapiens is merely one class among many, a subdivision of larger classes, linked to others in an evolutionary sequence. Race, scientifically speaking, is nothing more than one further subdivision, and the classification of races must rest solely on the same scientific basis as the remainder of the biological system. The criteria on which the classification of race is based must be (i) visible features; (ii) measurable; and (iii) inherited from a common ancestor: furthermore the anthropologist looks for the characteristics of a group, averaging the innumerable variations of individual people. An American anthropologist has defined race in this way: 'A race is a great subdivision of mankind, members of which, though varying individually, are characterised as a group by a certain combination of measurable features which have been derived from a common ancestor."

What are the features which the anthropologist measures in order to make his classification? The important ones include skin colour, stature, shape of head, face, nose, eye and type of hair.

1. *Skin colour.* We speak glibly of white, black and yellow races, and this is indeed a primary division of mankind. But it is also an over-simplification. Skin colour depends on a number of variables, such as the amount of pigment (melanin) in the skin, the depth of the blood capillaries under the skin, and the

thickness of the epidermis and the cuticle. Some pigment is almost always present (exceptional individuals who lack any pigment are called albino); but if the quantity is very small and the blood vessels can be seen underneath, then a pinkish white colour results. Sunbathing may intensify the melanin and bring it nearer the surface, giving a tanned appearance to so-called white people, but this is temporary. Little melanin and deep-seated blood vessels allows the parchment colour of the epidermis itself to be dominant, as among the so-called yellow races. Heavy pigmentation gives dark shades of brown.

2. *Stature is easily measured and classified into short, medium and tall.* With exceptions the male range is between about 130 cm/4 fit 3 in. and 200 cm/6 fit 7 in, the female range from 120 cm/4 fit 0 in to 187 cm/6 fit 2 in. Within limits food or the lack of food can effect stature, but it is nevertheless an inherited quality.
3. *The shape of the head, expressed as an index of breath over length* x *100 (the cephalic index), is a standard criterion.* Under 78.5 is considered a long head (dolichocephalic). 78.6 to 82.5 is medium (mesocephalic), more than 82.5 is broad (brachycephalic).
4. *The shape of face gives a variety of features.* It can be long or broad, the chin jutting out (prognathous) or receding (orthagnathous).
5. *The shape of the nose.* An index of the ratio of nose width at the nostrils to its length × 100 enables us to differentiate between long narrow noses (less than 70), medium nose (70-84) and short flat noses (over 84). Coupled with the long nose is the existence of a distinct bridge, whereas the broad nose is often depressed. Variations in shape are very considerable.
6. *The eye.* Eye colour can be classified in the same way as skin colour. But even more significant is the shape of the eye, for in this respect Mongoloids differ from other races. The upper fold of the Mongoloid eye droops over to give the impression of a silt-like opening. This is the epicanthic fold, and when it is more emphasised at the inner corner of the eye it tends to give the impression of an outward and upward slant, often accentuated because of the comparative absence of browridges and eyebrows in Mongoloids.

(7) *Hair.* Although there is a multitude of hair forms they can be conveniently grouped into three classes:

(a) straight hair, long and lank and rigid, and round in cross section;

(b) wavy hair; and

(c) kinky or woolly hair, much flatter in cross-section, and emerging from its follicle in a spiral, and in extreme cases forming hard tufts.

Colour, again depending on the amount of melanin in the hair, sometimes mixed with a red pigment, varies from ash blond (no melanin) and strawberry blond (much red pigment), to black (great amount of melanin).

These are some of the main features which are measured by the anthropologist. Even on these characteristics the number of possible variations must seem, at first glance, to be infinite. If one feature alone were used, then we would have very few faces; for instance skin colour would give three groups: Caucasoid (white), Mongoloid (yellow) and Negroid (black). But within any one of these, say Negroid, stature can vary from short to tall, and within either of these, head form can vary. Theoretically the number of combinations is immense. But in fact the number is cut down by the linkage of many of these characteristics, a linkage which suggests a common origin, and which is the only justification for attempting a classification at all. Woolly hair and black hair-colour go together, as do black skin and a broad nose: fair hair is usually wavy and goes with fair skin; and so on. It is the combination of measurable features which enables us to sort out a seeming mass of heterogeneous human beings into the divisions we call races: and these are always generalised for a group. Individual peculiarities must be forgotten—as indeed they are when we are viewing racial groups other then our own. To the European all Chinese look alike, because only the marked differences between European and Chinese are noticed.

The table which appears on the following page gives the main characteristics of the three great racial groups.

The three groups are known as primary races, but there are people who show mixtures of these basic characteristics which suggest that at some time after the establishment of the fundamental differences groups mixed and produced permanent strains with characteristics from two of the primary races. These are sometimes known as composite races. The Australian aborigines have a predominantly white base, of archaic type, mixed with Negroid: central Pacific peoples are mainly Mongoloid with mixture of Caucasoid, but the Papuan–Melanesian peoples are basically Negroid.

The Bushmen Hottentot are basically Negroid too, but mixed with very early Homo strains. Having disposed of this very broad outline of racial classes a geographer turns naturally to their distribution.

Main Characteristics of the World's Major Races

Trait	*Caucasoid*	*Mongoloid*	*Negroid*
Skin colour	Pale reddish white to olive-brown	Saffron to yellow-brown; some reddish brown	Brown to brown-black; some yellow-brown
Stature	Medium to all	Medium short to medium tall	Very short to tall
Head form	Long to broad and short; medium high to very high	Predominantly broad; medium high	Predominantly long; low to medium high
Face	Narrow to medium broad; no prognathism	Medium broad to very broad	Medium broad to narrow; strong prognathism
Nose	Usually high bridge; narrow to medium broad	Low to medium bridge; medium broad	Low to medium bridge; broad to very broad
Eyes	Light blue to dark brown; occasional lateral eye-fold	brown to dark brown; medial epicanthic fold common	Brown to brown-blak; vertical eye-fold common
Hair	Head: light blond to dark brown, fine to medium texture and straight to wavy Body: moderat to profuse	Head: brown to black, coarse texture and straight Body: sparse	Head: brown black to black, coarse texture and curly to woolly/frizzy
Body build	Linear to lateral; slender to rugged	Tends to lateral; some linearity	Tends to lateral and muscular; some linearity

What are called the indigenous, populations, though obviously this distribution is itself the result of prehistoric movement and mixing. The later historic spread of Caucasoid people, the peopling of the New World from Europe since the 17th century, would give much too confused a picture. The distribution of the three primary races is straightforward enough. Africa south of the Sahara is the home of the Negroids: the Caucasoids extend in a wide belt from north-west Europe to the Indian subcontinent, including northern Africa and parts of Asiatic Russia. Siberia east and south-east Asia and the Americas are Mongoloid. This distribution, together with our knowledge of racial movements, suggests that Homo spiens spread from an area within the Caucasoid belt, possibly somewhere vaguely in southern Russia: that in moving south and north-east, this early undifferentiated man was able to make two major adaptations to the environment, one to the hot wet climates of inter-tropical Africa where Negroid characteristics developed, and one to the cold climates of the north-east which produced the Mongoloids. The latter crossed a landbridge at the Bering Straits and peopled America at a comparatively late date (15,000 to 20,000 years ago). *i.e.* after the physical characteristics of the Mongoloids had become genetically stablised. There were, then, three major Old World adaptations to environment by a comparatively undifferentiated early Homo sapiens, together with an extension of one of these into the New World.

The distribution of the composite races is also interesting. They are peripheral to the supposed centre of dispersion. Some claim that a marginal location, sometimes coupled with isolation, would in itself promote changes: but it is also a fact that these marginal races show admixture, often of earlier strains of Homo which may have been less specialised. Some such mixture of Negroid and pre-Homo sapiens characteristics gave rise to the peculiarities of the Bushmen. Australian aborigines, with their paler Caucasoid features, make the link with Europe and the Middle East via the Dravidians of India. Negroids extended eastward to Papua, but by the time the Pacific is reached there are admixtures with Mongoloid and even with sub-branches of the Caucasoid races. Perhaps the most spectacular of the isolated groups is the Ainu of north Japan, a branch of the Caucasoid race; their location shows the possibilities of movement as well as the way in which racial features may be preserved far from the point of origin. The mixing of racial characteristics, the result of the free movement of man before as well as after the physical adaptations of his body to the earth's varied climate, is often preserved by the isolation of many of these peripheral areas. The picture looks a comparatively simple one. Of course it over-simplifies the facts. Each primary race has its subdivisions. The Negroids, for example, are divided into the Negro, the very tall Nilotic and the short Negrito. There are four main sub-races of the Caucasoids in Europe: Nordic, Mediterranean, Alpine and East Baltic.

Individuals belonging to the first are typically tall, slender, big-boned, with a fairly long and lofty head, a narrow face, prominent chin and narrow nose and lips. They are blue-eyed blonds. whose hair is fine and straight to wavy in type. They are found mainly in north European peoples, in Scandinavia (whee remnants of the classical type are found), lowland Britain, northern France, the Low Countries, northern Germany and Finland. Mediterranean peoples are shorter, long-headed, dark in colouring with dark eyes and black hair. Their name suggests their main distribution, but they also extend through Brittany to form a major element in western Britain. Indeed they form a basic racial substratum in the British Isles, but they were displaced and overlain by Nordic invaders in the east and south in historic times. Alpines are broad-headed, have broad squarish faces, concave noses and abundant dark wavy hair. They are medium in stature and stocky in build. They are found in a belt from France to the U.S.S.R., where they form a dominant element in the Slav population. The East Baltic group shows a mixture of Alpine and Nordic characteristics.

Social Bases of Racial Groups

It is essential to remember the limitations of maps of racial distribution. Two are especially important.

1. The world map suggests permanent zones. A moment's reflection on man's great mobility reminds one of how blurred these divisions are on the earth's surface. The distribution shown in the map is that of the late 15th century. Today Nordics, and to a lesser extent the other Caucasoid sub-races, have submerged the aboriginal peoples in practically the whole of North America, large tracts of South America, South Africa and Australasia. Within Europe, progressively increasing mobility has erased any racial boundaries that might once have existed. It is true to say that in any population one can only abstract racial characteristics from a large group of very heterogeneous individuals. Rarely are there whole populations in the Western world which retain common racial characteristics. Our definition of racial groups is based on average measurements.

2. This is because mobility has led to crossbreeding and this in its turn has made nonsense of any idea of a pure race. With every preceding generation the forebears of any individual multiply in a geometric progression. If a person today can trace his family back for eight generations, it means that his genetic constitution, and consequently his physical build is derived from 512 ancestors who lived about 1700: admixture is so probable, even in so short a span, that racial purity becomes a meaningless term.

It is small wonder that the layman is confused about the concept of race. The anthropological ideas we have discussed above are scientific but limited: the categories are to a large extent abstract and very difficult to apply to small groups of people. The anthropologist describes groups not individuals. Yet the layman has broadened the term race until it has become almost meaningless. He has, in addition, often ascribed to his own race a 'superiority'–implicitly a mental superiority–for which there is no evidence. These misconceptions have led to tragic chapters in man's history: small wonder that the very word 'race' has undesirable overtones.

But if the scientific concept of race tends to be an abstract, and if the distribution maps are so generalised, have racial characteristics no significance? Is the anthropological classification merely an academic exercise? Not entirely. The generalisations must be made: the group characteristics do exist: and they are relevant because the lay misconceptions are hung on these very hooks. Strange as it may seem, we must now turn to these lay misconceptions for it is on these that human interactions are based.

However diffident an anthropologist might be, even after extensive observations, to classify a person who has mixed racial characteristics, a layman will often make an unhesitating judgement. To some extent he must, because his behaviour in any given circumstance will depend upon such a snap judgement: it does not follow that his judgement will agree with the anthropologist's, but his is the one which will count in the circumstance. His classes are rigid. He thinks in terms of black and white. How does he deal with the in between cases, the racial mixtures? In the way in which he has been taught. If he comes from the southern states of the United States he will see any person with any observable Negroid features as black: indeed he may go further and fall back on known Negroid characteristics in the person's forebears to call an apparently white person black. Almost the reverse of this is true in South America, where a few observable Caucasoid features are enough for acceptance into the white group. From a layman's point of view race is what people have been taught it is: it is culturally defined. One American anthropologist puts in this way: 'A race is a group of people with more or less permanent distinguishing characteristics to which persons concerned attach certain interpretations.'

Certain physical features are used for rapid identification, and this is necessary to permit group cohesion. Man consciously groups himself, or classifies himself, but never objectively. His fundamental division of mankind is into two groups–'those who are like me, and those who are not'. 'Those who are not' are in the first place those who look different, so he attaches his label to a racial feature. The most obvious is colour, but stature, hair form and facial characteristics are all important. This is very far from the anthropologist's average measurements, but we are now dealing with the real bases of divisions, however far they may be from scientific truths, however dependent on error, misunderstanding and myth. For these half truths have been learned from previous generations and go back beyond the logical world of the scientist. Moreover, they are judgements and classifications which are often not physical and not inherited. An anthropologist has no evidence of differences in intelligence among racial groups; yet this is a supposed difference which apeals greatly to most peoples. 'We' are always mentally superior. Such values are cultural. In fact differences in intelligence are always very great between members of a single group, but difficult to assess between several groups. Intelligent, unintelligent, superior, inferior; all these are concepts about individuals, but they are spuriously attached to groups. Ethnocentricity, *i.e.* awareness of one's own ethnic (racial-cultural) group, emphasises the differences between groups, and often expresses it as antipathy to all other groups.

There is some excuse for this misleading, and often painful, characterisation of all groups other than one's own. It was suggested above that the layman's snap decision is based on some immediately apparent feature: he must be taught to do this, because he wishes his behaviour in a given situation to accord with that which has been taught to him, and because his chances of approaching the problem logically are infinitely remote. He has a shorthand method of classifying, not only racial appearance, but all the cultural elements of other ethnic groups. He has been taught that certain people dress in a certain way, talk peculiarly, have certain mannerisms. He has a ready-made picture of an individual from any other society. These 'stereotypes' are, of course, by their very nature caricatures, but they do mean (a) that communication is made easier and (b) that groups are easily if roughly classified. All English people carry a mental picture of a 'typical' Frenchman, a 'typical' Jew, a 'typical' American. However grossly misleading the images, passed on in a hundred and one unconscious ways to succeeding generations, they are the means of making a snap judgement: and they also save a lot of verbal description if one refers in passing to, for example, a 'Frenchman'.

Cultural Evolution of Man

The cultural elements–*i.e.* those ideas which have been passed on from one generation to the other–figure so largely in the kind of classification discussed above that we are justified in basing fundamental subdivision of mankind on these. It is more real to talk of ethnic groups than of racial groups, and although racial characteristics play a large part in determining ethnic groups, the latter are also based on cultural characteristics. Divisions in mankind can be most rigid when based on ways of living and thinking, and such divisions usually count for more than the broad racial classes.

One of the most important elements that binds people together is language. Difference in language is an immediate bar to communication between groups. Someone who 'speaks another language', even in the colloquial sense, is not one of 'us'. Language is the medium through which ideas are transmitted, so that one can expect a certain amount of homogeneity in the culture of common language groups. The world distribution of languages is very complex, for, although many of them come from common stocks, the similarities which linguistic scholars emphasise are lost on the individual groups who tend, if anything, to stress very minor differences. Differences even of dialect are enough to divide people. 'Speaking another language' can often signify even small nuances of meaning.

There are in Europe alone over thirty major languages, each the basis of an ethnic or cultural group. One need go no farther than the British Isles to see a considerable diversity in languages which was formerly even more marked. Nearly one person in three in Wales still speaks Welsh, and the language is the core of the separate identity of this group, the basis of a literature and the medium of certain ideas, traditions, etc., which are not shared with other groups. The remenants of Gaelic in Scotland are enough to remind us of one of the bases of separateness in the north, and few would deny that loss of language has completely obliterated these differences. Irish is still spoken in the western counties of Ireland and is the official state language. Manx has almost died out and Cornish is now fossilised in text-books. To some extent these Celtic languages can be grouped. There are basic similarities between Gaelic, Irish and Manx, the so-called q-Celts; and between Welsh and Cornish end Breton–the p-Celts. The people concerned are nevertheless distinct groups, and whether the language be alive, on the wane, or nearly extinct, one still recognises Welshmen, Scotsmen, Irishmen and Manxmen. Indeed, these groups have, or have had, separate political lives and their conquest by England was relatively late in their history. Ireland has re-emerged as a completely separate state: the Northern Irish, a mixture of Irish, Scots and English, have their own parliament, and so has the Isle of Man.

Religious beliefs are very deep-seated, sometimes uniting peoples of different racial and language groups, often dividing peoples who are otherwise identical. The earliest beliefs were animistic–*i.e.* they consisted of nature worship of one kind or another–and there are still many residual areas of this type of religion among the simpler peoples of the world. The origins of more advanced beliefs are linked with the great civilisations of the Old World and today have certain distribution patterns which suggest their diffusion.

Hinduism, very closely linked with Indian, is a compre-hensive religious system which embraces belief in many gods at one extreme and belief in one absolute being at the other; but the whole is welded together by certain fundamental attitudes such as worship of cattle, a doctrine of rebirth, and the caste system. The last makes rigid certain occupational differences, and was probably introduced to identify, first, groups of conquerors from the conquered, and later all the grades of division of labour. There is, then, a practical social element in Hinduism–one which further divides the society—as well as a system of belief and philosophy. The latter seemed so remote to Siddharttha Gautama (the Buddha), who lived in the 6th century B.C., that he founded a new religion which now embraces many millions in south-east Asia. His answer to the world's problems was a retreat into non-existence, and this may well seem attractive in a part of the world where existence is often at a miserably low level. China has two other religions, Confucianism and Taoism. The teachings of Confucius (551–478 B.C.) are concerned with social relationships, welded onto a more primitive system of beliefs which include ancestor worship. Lao Tze, a contemporary of Confucius, fastened on the spiritual and mystic elements in this primitive sub-stratum to teach a passive religion

(Taoism). Japan has its own system of beliefs, Shinto, basically a primitive animistic religion, given a deeper significance by the infusion of Buddhism in the 6th century A.D.

The other three great world religions. Judaism, Christianity and Islam, all arose in the Middle East, Christianity emerging from Judaism in what is now Israel, and Islam emerging in Arabia and owing something to both other religions. Judaism survived the break-up of the Jewish nation into minority sects in the cities of the Western world and the Middle East. Since the formation of the state of Israel it has re-emerged as a state religion almost at its point of origin. Both Christianity and Islam have spread remarkably. Christianity became a basic element in Western culture and spread into the New World with Europen overseas expansion, until its adherents now dominate four continents. The early spread of Islam was even more spectacular. Within a short time of the death of Mohammed (A.D. 632) his teachings had spread to the Indus and over the whole of North Africa, a main element in Arab civilisation. It spread temporarily into Spain, but in Europe today there are only a few remnant groups in the Balkans.

In more detail the divisions of Christianity and Islam reveal interesting distribution patterns. There is a sharp cleavage in Islam between Sunni and Shia, and the fact that the latter are dominant in Iran and in part of eastern Iraq has had important political consequences as a barrier to Arab unity. The division of Christianity into Roman Catholicism and Protestantism is equally marked. In Europe the former predominates in the Mediterranean countries, France and Ireland, and the latter predominates in Northern Europe. The Christian church has a third strong element derived from the eastern Mediterranean, that is the Greek Orthodox church which is predominant in south-east Europe and in Russia. Predominantly Roman Catholic South America and Protestant North America reflect stages in the conquest of those countries and the cultures of the peoples who settled them. Canada retains its very strong enclave of Roman Catholicism in Quebec, reflecting the religion of its 17th-century French settlers; and vast numbers of south and east European migrants have greatly increased, the proportion of Roman Catholics in north-eastern U.S.A.

It was stressed above that within the British Isles the Irish, Scots, Welsh and Manx have retained their identity because they originally had different languages and different cultures. But they are also groups which are identified with certain areas of the British Isles: Ireland, Scotland, Wales, Man. This is not necessarily true of all ethnic groups. At an earlier stage in man's history, when movement and instability were the rule rather than the exception, it was true of no ethnic group. The primary thing which distinguished such a group was social cohesion. This originated in ties of kinship: a person belonged to a certain group, irrespective of where they lived. A Jew was a Jew in any European country for those long centuries when he had no home of this own. But a some stage, probably in medieval times in Europe, a subtle change became apparent. As the historian Main put it–'England was once the country in which Englishmen lived: Englishmen are now the people who inhabit England.' There has been a transference of emphasis from social group to territory: whereas once a person was born into a specific society, now he is born within specific political boundaries.

This has far-reaching consequences. It brings us to a very recent stage, possibly the most advanced, in man's organisation. We began this chapter with a biological classification, then discussed social groups based on cultural differences; we must now deal with groups based on one element only in those cultural differences–the political. The classifications are not mutually exclusive: racial elements often underlie ethnic groups and ethnic differences underlie political differences: but there are also examples of several ethnic groups lying within one political division. But the last political grouping, with all the sanction of law and ideas of sovereignty, may well be the strongest of all.

Ethnic Groups

The subdivisions of mankind based on political organisation are the ones with which we are most familiar and are probably the most advanced and complex of all. They have by no means obliterated the other divisions, racial and ethnic, which have been dealt with, but what has often happened is that many ethnic problems have become subsidiary to state problems. Even problems of racial differences may well be contained within one state, although they may also often become world issues. The United States, although politically one unit, is deeply involved with all major racial groups, a fact which is emphasised in its own census with its primary division of the population into White, Black, Yellow and Red (even this classification, let alone its application, does not necessarily accord with anthropological classification): South Africa is similarly concerned with two major races. Australia is not less concerned to maintain a socalled 'racially pure' population. The United States is also concerned with numerous ethnic groups, for in the last 150 years every country in Europe has contributed to its very heterogeneous population. Very many countries have ethnic problems, not only because of the movement of people but because the drawing of new boundaries so often creates so-called 'minority groups'. Both racial groups and ethnic groups of this kind, which have no territory of their own but are merely parts of a larger, different population, are of great interest to anthropologists, sociologists and social geographers. The political geographer too must be aware of these problems and their consequences: not only because the location of such groups and their spatial relations are interesting but because of the part they play in the life of the state.

The attitude of a state towards other peoples and the degree to which it controls their movement affects the distribution of racial and ethnic groups. The White Australian Policy, aimed at the strictest control of people entering Australia in an effort to keep the population white, or even British, has a profound effect on the distribution of population in Occania and south-east Asia. For if this movement were unchecked there would undoubtedly be Asiatic migration and an extension of sout-heast Asian economy into Australia. The Australians look upon this as their main 'racial' problem for they can almost ignore the small number, about 50,000, of surviving aboriginals.

The United States faces the problems of ethnic as well as racial minority groups. Its own aboriginal peoples are now confined to reserves, their economies fossilised in the pages of text-books, they themselves museum specimens exploited by tourists and anthropologists. Their segregation is a measure of group antipathy basad on radical differences in cultural level. The United States also has a small number of Asiatics, particularly on the west coast, but their immigration was drastically regulated at the end of the last century. The negro population is the greatest 'alien' element, and the attitude of the white majority to this group is that of a rigid caste, based on social definitions attached to racial characteristics. The negro population is derived from a slave population, a cultural difference difficult to eradicate. A population of 750,000 in 1790 had given rise to 19 million by 1960. Until the latter part of the 19th century negroes spread with the extension of the cotton belt. Since emancipation there has been considerable movement to the north, particularly to the northern cities. But whereas a small number of negroes might lose themselves in a large city and provoke very little reaction, the building up of very large groups has resulted in the same restriction of distribution–*i.e,* segregation-in northern cities as they suffered in the south. In both New York and Chicago in particular, considerable number of negroes are crowded within narrow 'black' belts, giving a very distinctive distribution pattern.

The effort to maintain 'racial purity' is very different from the American attitude towards ethnic groups. These they try to assimilate: indeed, the United States is popularly looked upon as a 'melting pot'. In 1960 there were over 19 million people in the United States who had been born in another country, and 23,800,000 whose parents had been born outside the United States. At the end of the 18th century the vast mojority of white Americans could trace their ancestry to the British Isles. The stream of English, Welsh, Scots and Irish continued throughout the last century, augmented by millions of Germans, Scandinavians, French. Towards the end of the century this north-west European flood diminished. It was Italians, Poles and south-east European peoples who came in millions between 1890 and 1914. The 1914-18 war more or less put an end to immigration on this scale. These movements–which are dealt with in greater detail in another chapter–imply the break-away of millions of people from their parent ethnic group, and although some of them regroup in the United States and maintain their own language, religion, traditions, etc., the vast majority lose themselves in a larger social milieu: they are prepared to break theri home ties in exchange for the economic advantages of life in the United States. Some return home, of course, but the great majority make the transition, and even if first-generation immigrants–those who were born in another country–find the adaptation difficult, their children rarely fail to become fully-fledged Americans. The relative uniformity of the outward signs of being an American, together with the very rapid rate of change, makes adaptation fairly easy, and the Americans themselves, partly by their educational system, speed this process of Americanisation. The ethnic groups need not lose their identity completely and often add an indefinable quality and an enrichment to American life. The older groups have been absorbed the most and their distribution tends to be the same as the distribution of the population in general–with the exception of their absence in the southern states. But there are still interesting concentrations of Scandinavians in Wisconsin, for example, and a few north-east industrial towns have much more than their share of Welsh or German people. The newer immigrants are more confined to the north-east in general and to the cities in particular.

In time all these ethnic groups are bound to lose their identity almost entirely, for they have no internal cohesion strong enough to counter the attractions and advancements of American life.

This solution of absorption does not apply to the same degree north of the 49th parallel, where, within one federated state, there are two major ethnic groups–Canadians and French Canadians. The latter, derived almost entirely from an original French population of 60,000 which had settled in Quebec by the mid-17th century, are adamant on maintaining their cultural separateness. Often markedly anti-British and anti-American, and having had no ties with post-revolution France, they some-times express this by suggesting 'Latin' links, mainly religious and cultural, with Latin American states. In Quebec the imprint of a former French tenure system is clearly seen in the land-scape, in the long strip farms, in the street villages. French is still spoken and Roman Catholicism is the religion of the vast majority. Ottawa, federal capital, sits on the boundary of British Canada and French Canada, and, politically at least, study unites them.

Political Groups—The State

The political map of the world is one of the most familiar to students of geography. On the face of it the formation of states is a very acceptable division of the earth's surface and a convenient classification of its people, in spite of the fact that it is constantly changing. But further consideration shows the many problems which are involved in this political subdivision. To begin with, although a state boundary line is so very easy to delimit on the map, it is much more difficult to demarcate on the ground: indeed most of those precisely drawn lines on the political map of the world have never been demarcated. It is even misleading to think of the divisions as being lines. The 'line' is much more often a zone, or a frontier. At first glance an island may seem to be an ideal unit for one state. But even a coast-line is really a zone. Not only is there a small intertidal zone, not only is there sometimes the no-man's-land of quarantine, or an Ellis Island, but territorial sovereignty extends beyond the land, and over the sea; formerly to fairly well-accepted limits of three nautical miles, but now over very varying and indeterminate distances. In the past states have dominated for beyond even this. The Mediterranean was Mare Nostrum to the Romans, Venice collected dues from all who used the Adriatic, and Tudor England claimed sovereignty of the Narrow Seas from Norway to Spain.

To return to land-locked states, the geographer is interested to see to what extent physical features are related to frontiers and boundaries. If mountains are high enough they make good frontiers because they are unpopulated regions and make good frontiers because they are unpopulated regions and less likely to be bones of contention. The Himalayas, the Alps and the Pyrenees, all have major frontiers along them. Yet their are pitfalls in thinking that all mountains make good frontiers. Lower ranges may well be the homes of mountain communities: even in the Andes, large communities of people live at very high altitudes. And even in the high ranges the boundary becomes important at passes, which are low and vital to trade.

When new countries were first opened up to European expansion, rivers seemed to be very acceptable boundaries: they were easily recognised, and involved no demarcation. Such boundaries are commonly found in many newly settled countries. But their relative absence in Europe is significant, for hundreds of years of adjustment, by wars and treaties, have shown the river to be a poor boundary for many reasons. In a developed country a river is usually the centre of a social unit, a valley, and to split such a unit is unwise. In the United States, where the Mississippi and Missouri separate several states there are often towns on either bank, *e.g.,* the two St. Louis, two Kansas Cities, Council Bluffs and Omaha. It is only the federal union of all these states which makes these river boundaries acceptable. An international boundary on a river can be critical, particularly when its water is used *e.g.,* some of the Punjab rivers: and the use of the river by transport is easier if controlled by one state.

Physically a river can play tricks on men's efforts at state demarcation. A mature river like the Rio Grande, dividing the United States and Mexico, is constantly changing course. Oxbow lakes are cut off leaving large tracts of territory and communities of people-in another country. This has happened so often that these two countries have agreed on certain legal readjustments so that the state boundary is no longer entirely along the line of the river.

One of the Most common boundary lines in new countries is the straight line. Such geometric boundaries are particularly useful in so-called 'empty lands', and are often retained after the country has been settled. They are lines of latitude and longitude, and are particularly promenent in North America and Australia. The most famous is the 49th parallel, the dominating east-west boundary between the United States and Canada. This is the easiest kind of line to delimit, though not always so easy to demarcate. A similar line running across a very densely populated country, as in Korea, leads to considerable confusion and hardship.

There are the most common types of boundaries. Their classification and evolution are of interest to the political geographer:

their delimitation is vital to society. For in those parts of the earth which are densely populated, like Europe, and where constant movement has blurred ethnic distinctions, it is very difficult indeed to define boundaries which do not conflict with the interest of some groups or other. Minorities are often created by boundary delimitation, *i.e.,* pockets of one ethnic or language group are caught within the national boundaries of another, usually because strategic and economic considerations are uppermost in boundary-making. Tension is inevitable. The shift in allegiance of Alsace and Lorraine to Germany in 1871, and back to France after the First World War, had a great effect on the society, made worse by the fact that the population was already a mixed one. An Austrian minority in the Italian Tyrol, the result of a transfer of territory after the First World War, has created tension since. Such problems are legion in Eastern Europe where boundaries have been reshaped several times in this century. Indeed this region of instability is often called the 'shatter zone', for boundary changes, minority problems, and economic and political instability often go together.

The ideal would be to have a homogeneous population within each state boundary: in reality most boundaries are strategically determined and ignore all social facts. The political map of India has undergone remarkable changes since 1947, moving towards on ideal, even if imperfectly. Under British domination India was a vast collection of states, some small, some very large. Britain exercised direct control over many of these, which together made up British India: with the remainder she had treaties defining her relationship with the native princes who kept a certain degree of sovereignty. In this division into British India and the princely states, the former was determined strategically, *i.e.,* with few exceptions Britain directly controlled the coast, the two great river valleys, and the north-west frontier; and ensured that this directly controlled territory split up the major native states in such a way that they had no contact one with the other. In 1947 India was partitioned in a very different way. The basis was religious, for the new state of Pakistan was based on Moslem majorities. This, incidentally, gave rise to a Western and an Eastern Pakistan, separated from each other by India. This partition was a realistic appraisal of a vital difference in two cultures, expressed as religion. In 1958 India took the next logical step and rearranged her own internal state boundaries on the basis of language. This was difficult, and, as with the religious solution, it involved creating some minorities, many of them discountered. But the modern political pattern in India is now a logical one based on cultural variations: imperfect though it is it reflects Indian society and culture. It is very different from the imposed British pattern which made no concession to these differences.

Within its international boundary every state exercises paramount authority by law. This is the sovereignty of states which theoretically makes them all equal. But a geographer is not greatly concerned with this equality. To him it is variety of states–in size, population, economic wealth—which is the most interesting feature of the political map. He is also interested in their origins, growth and expansion.

Most states have within them a nucleus, a centre of manpower and wealth; this is often the area from which, in former times, control has spread to the present boundaries of the state. In Britain, Lowland England, and in particular the London Basin, was such a nucleus. The Peris Basin, similarly, has become dominant in France. Associated with the rise of the nation state was the rise of the capital city, for the wealth of the nuclear region was associated with the growth of the capital city which was almost always its centre. Just as the strength of the state lay in the nucleus, so its wealth and influence became concentrated in the capital. The 16th and 17th centuries saw the beginning of this disproportionate growth of the capital city which is still apparent.

In a way every capital city is unique, but many have points in common which have led to attempts at classification. It is certainly useful to note that besides those, like London and Paris, which lie within the rich lands which gave birth to them, there are capitals which lie outside such nuclear areas. A capital until recently, Istanbul, or Constantinople, had no rich hinterland: in its heyday even its food came from Egypt and the Black Sea. Copenhagen is peripheral in Denmark. In both cases the city's growth was related to former extensions of its territories. Constantinople was the centre of Byzantium: it was a supra-national city: it defined with the decline of the Ottoman Empire, and when Turkey emerged after the First World War as a new state its capital was shifted to Ankara. Similarly, Copenhagen was once the centre of a state that dominated the Baltic Sea.

Other capital cities mark points of contact between societies, or sites where invaders have entered new lands. This is particularly so in the New World where urban civilisation was almost always brought by Europeans. The state capitals of Australian-Darwin, Brisbane, Sydney, Melbourne, Adelaide, Perth, Hobart–are all on or near the coast, the toeholds of Europeans in a strange land. Bombay, Madras and Calcutta show the dependence on the alien British on naval strength. The early capitals of North America were coastal, though they shifted inland when settlement went beyond the Alleghenies. In South America there is an interesting contrast between the Andean capitals, many above 2,750 m/9,000 it, which lie in rich basins and were first founded by pre-Columban cultures like the Inca, and the later European toeholds which developed into Buenos Aires, Montevideo and Rio de Janeiro.

Lastly some capitals can be classified as artificial in the sense that they were built purely from political considerations, have little economic base, and have therefore no direct relationship with a state nucleus. Madrid was probably the first of such capitlas, its central position a conscious attempt at pulling together the disparate richer regions of Spain which were peripheral. And so, in the 15th century, a lonely outpost on an unattractive and barren plateau gradually began to be transformed into a capital city.

The outstanding example of the 18th century is Washington. Washington lay roughly mid-way between northern and southern elements when the newly formed United States was still an Atlantic community. Today it is a good example of the single-purpose city, for half its employed workers are engaged in government, the other half in services.

The same desire to balance various political attraction led to the choice of a lonely bush station as the site of Canberra at the beginning of this century. This again is a city which has only one purpose, and whose only justification is federal government. The most recent and exciting artificial capital is Brasilia. Placed well outside the settled areas of Brazil, it is hoped that it will not only be a symbol of Brazil's maturity–the cutting of the link with the coast and thus with the European heritage–but that its central position will be a focus for the several states of Brazil and a growing point for future development.

Grouping of States

Washington, Canberra and Brasilia are all federal capitals, a reminder that many states are grouped together into larger entities. The merging of status on an equal footing is a federation. The extension of the authority of an expanding state over other countries, by war or by diplomacy, leads to empires, the different parts of which will have varying degrees of sovereignty. From the disintegration of former empires, new states are continually being formed. Political geography is a dynamic subject, and grouping, disruption and regrouping and constantly taking place. Three of the most extensive and powerful groupings are the British Empire, the United States of America and the Union of Soviet Socialist Republics. Geographically there is much in common between the two latter, and they contrast markedly with the first. Both the U.S.A. and the U.S.S.R. are continental, *i.e.,* their hegemony extends almost entirely over a single land mass.

The United States expanded by acquisition of territory in the first half of the 19th century from its Atlantic seaboard origins to the Pacific. The U.S.S.R. expanded from its European core into Siberia and eastward to the Pacific. (Their interests overlapped in Alaska, which the United States bought from Russia in 1867.) The integration of these great land masses depended mainly on communications and, in both, transcontinental railway lines were vital links. In great constant British territories overseas are very scattered. The links depended this time on the steamship. Shipping routes became vital to Britain, and so did depots and strategic defensive points along them.

The furthermost parts of the empire are linked by a great number of islands; some very small, but all essential in holding together the empire during its formative period. British overseas possessions are complex in another way. Some of the territories were formerly comparatively empty lands in temperate zones which could be settled by British people. These have developed modes of life similar to those in the mother country, and have attained complete sovereignty: they are the Commonwealth countries, Canads, Australia and New Zealand. Other territories lay in the tropics, already thickly peopled, but exploited by comparatively few white people for exotic raw materials. Such colonies and protectorates, the majority in Africa, are now emerging as new states.

Compared with the diversity of the British overseas possessions, the states of the U.S.A. are remarkably uniform; because in their organisation and basic culture they are all derived from a common stem. On the other hand, the U.S.S.R. covers an amazing complexity which is reflected in its organisation. Ideally its republics are based on major ethnic groups, but the largest republics also have within them territories–usually much less developed–which have less autonomy. The whole is also linked in one intricate economic system.

So-called empires or federations are not the only kind of political grouping, though they have the greatest cohesion. Military and economic ties can be entered into by any group of states which may then identify themselves as having common aims.

With the exception of Canada, the states of the Western hemisphere have certain interests in common which they discuss as members of the Pan-American Union. N.A.T.O. is a military/economic union which spans the Atlantic. Inside Europe there have been marked trends towards common economic aims in the European Steel Community, the European Common Market and the European Free Trade Area.

These alliances constantly change and remind us that territorial expansion is not the only way in which one state may influence another. A political geographer should be aware of economic and ideological influences which tend to break down the seemingly rigid boundaries of the political map.

4

Population Distribution

Unit quite recently the systematic study of population had been largely neglected by geographers, in contrast with other fields of human geography such as agriculture, industry and settlement which have a long-established tradition of systematic analysis.

However, in recent years there has been a growing awareness of the importance of population studies within the broad framework of human geography. Population geography has rapidly advanced from a peripheral position within the discipline to the stage where it has been claimed that 'numbers' densities and qualities of the population provide an essential background for all geography.

Population is the point of reference from which all the other elements are observed and from which they all, singly and collectively, derive significance and meaning. It is population which furnishes the focus.' (G.T. Trewartha)

Population geography is concerned with the study of demographic processes and their consequences in an environmental context. It may thus be distinguished from demography by its emphasis on the spatial variations in the growth, movement and composition of populations, and its concern with the social and economic implications of these variations.

The development of the subject has been severely limited by a lack of demographic data for many parts of the world, but, like other branches of geography, it is concerned with description, analysis and explanation, although a large part of the work to date has been concerned with population mapping and descriptive studies, simply in order to establish the data base which must precede analysis.

Population Data

One of the most difficult problems facing the population geographer is the varied quality of population data available for different countries and regions of the world. In general, the economically advanced nations have more accurate and detailed statistics than the developing countries, although in recent year the United Nations Organisation (UNO) has assisted many countries with the vital process of demographic data collection. The publications of the UNO have the advantage of compiling and collating dispersed and fragmentary material on a world basis. Thus, the reports and studies of the UNO, FAO, UNESCO and WHO provide vital documents for many aspects of population study.

At national level there are two main sources of population data: census reports and various registers of population. The latter involve the compulsory and legal registration of vital events such as births, deaths, marriages, divorces and adoptions.

In most countries such registers have a far longer history than the national census, and it is generally agreed that national registers are more accurate and reliable than census returns. In addition, there are many specialised sources of information dealing with specific aspects of population. In the UK, for example, the Registrar-General's Quarterly Returns, reports of the Departments of Health and Social Security and Employment, the Central Statistical Office, Regional Health Authorities and so on, all provide useful information about various specific aspects of population.

A census has been defined as 'the total process of collecting, compiling, evaluation, analysing and publishing demographic, economic and social data pertaining, at a specific time, to all persons in a country or in a well-defined part of a country' (UNO, 1967). Although crude attempts were made to enumerate population in early times, modern census recording covers only a relatively recent time-span.

The earliest modern censuses were organised in Scandinavia—in Sweden in 1748 and in Norway and Denmark in 1769. In fact, the Scandinavia countries still maintain a leading position in the collection of population statistics that are noted for their accuracy and detail. By contrast, the initial, though very imperfect, census of the USA was held in 1790, while it was not until 1801 that Britain held its first census.

Problems of Using Data Sources

Contrary to popular opinion, census statistics are subject to many errors and limitations. These result from omissions and double entries, either as a result of genuine error, illiteracy or with the intention to deceive for fiscal or political motives. National registers, while generally more reliable, also suffer from inaccuracies. In remote rural areas infants who die before registration may not be enumerated. It has been estimated, for example, that even in the USA almost 2 percent of births still escape registration.

For the geographer working on census statistics there are many problems. Some of these result from breaks in sequence or the lack of a regular sequence in the census years. In the UK, for example, there was no census in 1941, while in France the rhythm of a supposedly five-yearly census interval was broken after the Second World War when enumeration was carried out in 1946, 1954, 1962, 1968, 1975 and 1982. Other states make no attempt at a regular census interval. Russia, for example, has held a census on six occasions only, in 1879, 1926 1939, 1959,1970 and 1979. In China, the first census was not held until as late as 1953, when almost one fifth of the world's population was brought into official statistics at a single stroke. Clearly, international comparisons of population statistics are made extremely difficult by such variations in census frequency.

It should also be appreciated that different countries employ different methods of data collection. In the UK, for example, individuals are recorded according to the place where they are found at the time of the census, whereas in the USA individuals are recorded according to their usual place of residence.

However, a more serious problem is the lack of uniformity in the categories of data collected and the systems of classification employed in the published statistics. Various divisions are used for the classification of population into age groups; definitions of urban and rural population vary widely from country to country; definitions of occupational groups are similarly varied; migration statistics are notoriously unreliable, and lack international comparability owing to diverse methods of compilation and a lack of uniformity in the definition of temporary and permanent migrants. For some years the UN Statistical Office has sought to remedy these defects by encouraging countries to take their census in the same years and to follow an international scheme of data classification. Nevertheless, the problems remain. For many parts of the world no reliable demographic data whatsoever are available, and even in the more statistically advanced countries the amount of uncertainty is still considerable.

A further problem must be mentioned at this point. It will be apparent that the geographer working on census data is dependent on the size and shape of the areas for which the enumerations have been made. Frequently these censal divisions are based on administrative or even arbitrary boundaries which have little or no geographical significance. Furthermore, these censal units are subject to frequent changes and adjustments, thus invalidating comparisons of data over a long period of time. In this connection a proposal to replace the present irregular census divisions in the UK by a series of 100 m grid squares would be of great benefit to the population geographer. This would not only give statistical stability to the data base, but would also facilitate the computer processing of results.

Influences on Population Distribution

Environmental factors such as high altitude, extreme cold and aridity are the essential influences upon population distribution. However, closer consideration shows that such physical factors provide no more than a partial and determini-stic explanation. As described in Chapter One, any attempt to define a climatic or environmental optimum for human habitation proves very difficult. Furthermore, man is by no means passive in his choice of areas for settlement and everywhere demonstrates an ability to exercise some control over his environment. Thus, the analysis of any population distribution, whether on a local, regional or even world scale, must inevitably take into account various social, demographic, economic, political and historical factors as well as purely physical influences.

If should also be emphasised that the factors affecting population distribution rarely, if ever, operate in isolation, but rather in combination. Thus, it is virtually impossible to evaluate the level of influence of any single factor. Typically, a whole range of influences is relevant, but their interplay is generally extremely complex. This is especially true of the economically advanced nations of the world. Indeed, it has been suggested that the role of physical factors in the spatial distribution of population declines in direct importance as civilisation advances in complexity.

Physical Influences on Population Distribution

It is obvious that population numbers and densities decrease with altitude in response to the increasing difficulties involved in the settlement of high-level environments. High altitude imposes physiological limits upon human existence through reduced atmospheric pressure and low oxygen content. Consequently very few permanent settlements are found in the Andes and Himalayas above 5, 500 m, while La Paz, the capital of Bolivia, at a height of 3,640 m, is the highest city in the world.

The influence of altitude cannot, of course, be divorced from that of latitude. In low latitudes, high-plateau areas provide positive advantages and are often favoured for settlement on account of the climatic amelioration that they provide. Conversely, in high latitudes, areas of low ground are generally sought out by the scanty population of such areas. It is thus possible to refer to 'mountains that repel and mountains that attract'.

Similarly, altitude cannot be disassociated from the factor of relief as an influence upon population distribution. Steep gradients, exposure and rugged terrain all tend to deter settlement by restricting movement and the possibilities of cultivation. Thus, an examination of population distribution maps will show that some of the most abrupt changes in population density occur at the junction of mountain and plain. However, within upland areas the pattern of valleys may have important effects. Certain valleys provide lines of penetration and favour communications and settlement, while cul-de-sac valleys remain isolated and sparsely populated. Similarly, on the more densely populated plains, rivers may exert either a positive or a negative effect. Most attract settlement, especially at crossing points, confluences and the head of navigation, but others may be liable to flooding and deter settlement.

Reference has already been made to the influence of climate upon population distribution and its effects on human activity (Chapter One). Such influences are exteremely difficult to evaluate. Extremes of heat, cold, humidity and aridity all clearly deter settlement, but climatic optima are very difficult to define. It is of interest to note that two of the world's primary concentrations of population lie in middle latitudes, while the third is located mainly in the tropics. The difficulties of interpreting climatic influences may be further illustrated by the fact that virtually indentical climatic environments support vastly different population densities. For example, the island of Java and the Amazon Basin both experience an equatorial type of climate, but whereas Java has a population density of over 500 persons per square kilometre the Amazon Basin is one of the most sparsely populated regions of the world with an average density of less than 1 person per square kilometre.

In a similar manner, the influence of soils on population distribution is undeniably important but difficult to define as an isolated factor. Deltaic and alluvial soils frequently attract agricultural population, while podzols and laterites, with their limited possibilities for cultivaton, generally support only sparse populations. The attractiveness of particular soil types also depends to a large extent upon the agricultural technology of the population in question. Settlement of the American Prairies in the mid-nineteenth century, for example, was greatly facilitated by the development of the steel plough, the windmill and barbed wire.

World patterns of climate and soils strongly influence the distribution of major vegetation types, which in turn provide constrasting environments for a variety of agricultural activities. These are often associated with particular levels of population density. Consider, for example, the contrasting population densities of the ranching lands of the New World and areas of Asian rice cultivation. However, no firm rules can be established concerning the relationships between type of agricultural economy and population density.

The contemporary distribution of population in many parts of the world is also influenced to a large extent by the location of mineral and energy resources. For example, the population map of Western Europe reflects the distribution of coalfields and their associated industrial conurbations. In several instances these old centres of mining support densities of over 1,000 persons per square kilometre. The South African Rand, the Appalachian coalfields, the Donetz Basin and many other areas provide examples of population concentrations associated with the working of local mineral deposits. In the case of northern Canada and interior Australia the presence of minerals has attracted small settlements far beyond the limits of the ecumene. It will be appreciated that the influence of mineral and energy resources upon population depends on a whole range of related factors such as market demand, capital for development, availability of labour supply, presence of transportation links and cost of production. Indeed, all the physical influences outlined above should be evaluated in relation to the economic, social and political condition in the area concerned.

Economics, Social and Political Influences on Population Distribution

It has already been implied that the density of population in a particular area depends to a large extent upon the type and scale of economic activity in that area. Technological and economic advances are usually associated with changes in population density and distribution. For example, the North American Prairies presented different opportunities for the Indians with their hunting economy, the nineteenth-century ranchers, the later settled agriculturalists and the modern industrialised and largely urbanised society. Each stage in the economic development of the region involved a growth of numbers and substantial changes in the distribution pattern. The Industrial Revolution has been a particularly potent force of change in many countries of the world. Pre-industrial agricultural populations, often fairly evenly distributed, were attracted to sources of energy, particularly coalfields, as well as lines of transport and communications and ports. Dense population concentrations replaced long-established patterns of dispersal and generally even distribution. On the whole, it is true to say that increasing complexity and diversity of economic activity encourages unevenness of population distribution.

In the present century, with increasing government control over economic activity, political influences have emerged as a significant factor affecting population patterns. In Communist countries, population may be directed to areas of social or economic need, while in the Western World various inducements may be offered to encourage or assist migration to new towns, development areas or simply away from overcrowded conurbations. Political events have also been responsible throughout history for mass-migrations of population. The post-war movement of refugees into West Germany or the enforced expulsion of Asians from Uganda in 1972 provide examples of such a process.

Finally, it will be evident that historical processes must also be taken into account in any analysis of contemporary population patterns. The duration of settlement in any area is of fundamental importance. The relatively recent settlement of Australia is a basis reason for its low population density (2 persons per square kilometre), while the high density of India (216 persons per square kilometre) may be partly explained in terms of its long history of civilisation and occupancy. Nevertheless, it cannot be concluded that the highest population densities are found in the longest-settled regions. Numerous examples of formerly prosperous and densely populated regions, now only sparsely populated—such as parts of North Africa and Mesopotamia, the Yucatan Peninsula and eastern Sri Lanka—have led many historians and others to propose the idea of cycles of occupation, whereby population numbers and densities increase and then decline, only to be followed by a second cycle of population growth.

Measures of Population Density and Distribution

So far in this chapter the terms density and distribution have been used without definition, but, in fact, they do have precise and different meanings. Distribution refers so the actual pattern of spacing of units or individuals; density, on the other hand, is an expression of the ratio between total population and land area.

Measures of Population Density

Crude population density is a measure of the average number of individuals per unit area. For example, the UK has an average population density of 232 persons per square kilometre. As an average figure it suffers from all the limitations that this implies. A crude population density figure provides no information about extreme values within a territory, and comparisons of density figures are meaningful only for small area units such as parishes and communes, but not for large units such as nations or continents. A crude density figure alone can never be used as an index of overpopulation.

Because of these limitations, various refinements of crude population density are sometimes employed. For example, densities can be calculated for inhabited or cultivated areas only rather than for 'gross' area. The latter is known as a physiolo-gical or nutritional density. Conversely, the density of certain sectors of the population can be calculated against total area. For example, it may be of interest to have a density figure for the industrial or agrarian sections of the population. The latter is referred to as an agricultural density. In urban areas where high-rise blocks of flats invalidate simple relationships between population and area, room density, or average number of persons per room, provides an index widely used by planners and sociologists.

Measures of Centrality, Dispersion and Concentration of Populations

Various techniques have been developed to express the central tendency of populations. Most have been concerned with the establishment of a theoretical central point within a given distribution.

The mean centre of a distribution is 'the point about which distances measured to all individuals of a population will, when squared, add up to a minimum value'. For grouped data such as the states or counties within a national territory it may be expressed as $\Sigma(pr^2)$, where p represents the population within a unit of area and r is the distance between the mid-point of the unit and the mean centre. Time-series maps showing changes in the position of the mean centre can provide an indication of the directional shifts in a changing pattern of population distribution. For example, the mean centre of the population of the USA has shown a steady shift westwards along the 39th parallel since 1790.

The median centre, or centre of convergence, is defined as 'the point where the whole population could be assembled with the minimum aggregate travel distance'. Using the substitutions noted above it may be expressed for grouped data as S(*pr*). As with the mean centre, calculations of the location of the median centre is difficult and tedious, but at a regional scale it does have a practical application in the determination of optimum locations for centralised services such as schools, hospitals and hypermarkets.

The modal centre simply describes the point or area of maximum surface density within a distribution. Most countries are uni-modal; for example, London is the modal centre of the U.K. A few countries are bi-modal, such as Canada with Montreal and Toronto, and Australia with Sydney and Melbourne, while a very small number of countries are multi-modal; for example, India with Bombay, Calcutta, Delhi and Madras.

When the location of these theoretical centres has been established, various statistical methods may be employed to analyse the extent to which population is dispersed around them. Standard distance deviation may be compared with the standard deviation of linear statistics, and is found by dividing the mean centre of population, $\Sigma(pr^2)$, by the total population, P, and then taking the square root of the resulting mean square

$$Sr = \sqrt{\frac{\Sigma(pr^2)}{P}}$$

Mean distance deviation is likewise the counterpart of mean deviation for linear statistics. It provides a measure of the mean distance of individuals from the median centre of population. It is found by dividing the median centre, $\Sigma(pr)$, by the total population P

$$md_r = \frac{\Sigma(pr)}{P}$$

A point of particular interest in many geographical studies is the unevenness or degree of concentration of populations. Concentration is at a maximum in the hypothetical situation in which the total population is assembled at one point, and is at a minimum where each individual is located at an equal distance from his neighbour. The tendency of a population distribution towards one or other of these two hypothetical extremes can be measured by means of a graphical device known as a Lorenz Curve. This involves plotting cumulative percentages of population against cumulative percentages of area. Changes in the form of the graphical curve over time indicated increasing or decreasing concentration of population.

World Distribution of Population

If a glance is set on the population distribution map 20-1 of the world, we will find that the population of the earth is very unevenly distributed. One fourth of the land surface holds approximately 90% of the total population of the world. The remaining 10 percent is very thinly distributed over 75% area of the earth, most of which is too cold or arid for agriculture.

The four principal regions of dense population on the earth are :

1. *Eastern and Southern Asia*—including India, China and Japan, which contains over half the world's total population.
2. *Western and Central Europe*—a fourth of all population, 500 million, occupies an area which amount to roughly 6% of all land.
3. *South Eastern Canada*—Majority Canadian settlements were within 1000 kilometres of Montreal, and only a few were as far away as 1500 km.
4. *North-Eastern United States*—Each country of the South-eastern Canada and North eastern United States of North America has a very patchy population distribution with the bulk of its populations localized on a small proportion of its total area.

These four regions, with only one tenth of the land surface of the globe sustain two thirds of the earth's people. In all these four zones the density per square kilometre exceeds 300, but there are many local variations. In Yangtze valley the density is 700 persons per square kilometre, in Hwang Ho valley it is 600; and about 400 per square kilometre in the Ganga delta. The persent unprecedented increase in human numbers is not a process which is going on equally all over the world. In spite of excessive density of population in certain regions about half the total area of the earth is now empty space. In certain regions the density is below 2 persons per km^2. These are : Hot deserts, Equatorial regions, Mountains and Plateaus of Central Asia and Sub-polar areas. The problem of scarcity in over populated regions can best be solved through migration of people to open spaces.

Population Distribution in Asia

The distribution of population in Asia is unique unlike the distribution in Europe, it is unevenly and unequally distributed. The great human agglomerations in Asia cease at about the same latitude where they begin in Europe. According to Blache, "the Asiatic agglomerations were created and have grown up under the influence of one dominant cause, that of the Monsoons. Centres of dense population—scattered at first—have generally approached one another and finally merged,.......the various contributing cause are sunlight, rivers and rainfall, all of which relentlessly over-stimulate the productive soil". Some parts of Asia have densest population such as Java, whereas in other parts like Tibet are very thinly populated. In Siberia the density is fourteen persons per km^2, is generally known as the "Store house of the future". In Middle East some parts in Arabia have only one person per km^2 and have many potentialities for future growth.

Asia in its one-third portion contains three fourths of its total population. All the plains of Asia are densely populated. Such contracts in density is not experienced in any other continent. There are certain reasons of this unequal distribution of population and the first cause of all is that the continent of Asia is agricultural. Therefore, the people concentrate only in those parts where the land is fertile. All the river plains where the soil is fertile, irrigation facilities are available, means of communication are developed, are densely populated. The Indo-Ganga plain of India, Sikiang Yangtze Kiang in Peoples Republic of China, and Kwanto plain in Japan are densely populated regions of Asia. All the eastern half of Asia is drained by rivers which flow towards South and East. All these plains tend to make this South-Eastern part of continent is the most densely populated portion in Asia, and probably in the whole world.

The most important ecological factor which determines the distribution of population is climate. Those parts of Asia where the rainfall is enough for the growth of grains are inhabited by people. But on the contrary, where the rain fall is less than 50 cm. these parts are sparsely populated. Such parts are in the interior of the continent. Mongolia, Turkistan, Central Siberia, Northern Manchuria and Northern Japan are such regions where the population is found very sparse. In the South East Asia Monsoon brust with their utmost capacity and greater part of their moisture is downpoures here. The rainfall is one of the most important ecological factor in determining the distribution of population. In western or Middle-East Asia, Arabia, Iran, Iraq has rainfall below 25 cm. and so that the population is negligible. The Middle East is often taken as the desert because the major portion is arid. But the valley or Euphrates and Tigris are again a populated area.

The next ecological factor is physical features. The high plateau and mountainous regions are left desolated in spite of the mineral resources. It is only due to the lack of communication. The Shan plateau, the mountains of China, the Pegu Yoma and Arakan Yoma are sparsely populated though there are mineral resources. The last factor is that there is lack of mobility in Asiatic peoples. People do not wise to go to some other land leaving their parental land. They are most conservative and believe in idol worship. Chinese say "who will worship the graves of our ancestors" if they go away to some other lands.

So these are some factors that lead to the unequal distribution of population. Some parts are over populated and some are underpopulated. With this unequal distribution there arise several problems. Under-populated areas are very backward. The people are few and due to lack of labour no progress is possible. In Tibet there are no means of communication, no industries, neither any social development nor economic development. The people of Shan plateau are still in primitive stage. In central and in some western part of Asia there pervades monotony. There is no vitality in the life of people. People of Central Asia are nomades. They have less economic activities. Natural resources are not used to the utmost capacity. And to use them it often happens so that foreigners enter and established their industries there to explore the natural resources. As Tibet is exploited by Chinese. In middle-East due to lack of technicians and technical know how, the mineral oil is explored by foreigners. Foreigners make their holding more and more strong day by day, when they settled in an underpopulated country. Tibet and Mongolia have been exploited by China only due to the underpopulation. They had no military force. On the contrary, China is a nation of over 1,500 millions (1955) people has greater man power than any other country.

Iraq is small country processing three vital resources—oil, water and land. Generally is known as a country of retarded development. In 1951 the Anglo-Iranian oil company was nationalised by the Iranian government. Britain withdrew all technicians and experts, the result being that the production of oil gone under an abrupt reduction from 129 million barrels to 9 million barrels. Complexity of geographical conditions and interplay of widely differing influences explain the anamalies and problems of population in the Middle East. South-west Asia is a bridge and a barrier between East and west, so that the positional factors are of great importance and, with the oil resources of these lands of backward technique, are responsible for the conflict of interests of great external powers of this area—national interest which restrict real independence of the several small states into which is divided. There are resources and the policies of nations their own.

Turkey is a miniature continent and its population is mostly distributed along the rims of Anatolian plateau may be described as a province of transition between Asia and Europe. No country can make any industrial progress until and unless there is cheap and efficient labour and resources. But overpopulation is also a curse for a country. China is a vast country. If occupies about 9.9 million sq. kilometres. It stretches from east to west for more than 4000 kilometres and only a little than this distance from north to south. The population amounts to something between 600 and 700 million. Three fourths of the area consists of mountain ranges, which are hardly, it at all, suitable for agriculture, and here the problems of soil erosion are acute. Millions of hectares of mountainous areas have become eroded to such an extent as to be no longer cultivated. The area subject to erosion aggregates over 1,500,000 km^2. Another large part of China is occupied by virtually lifeless deserts.

The principal agricultural regions and the larger part of the population are confined to the deltaic alluvial plains along the Pacific coast. Here is the world's highest population density, commonly reaching 3000-4000 people per square kilometre, and in some places ever more than 500 persons per square km.

There are a number of reasons why the population has not overflowed into the sparsely settled lands that surround the coastal and central provineces of China. Nature provides the most important reason. The settled areas are singed by regions of non-productive soils with either insufficient rainfall or unsuitable topography or both, areas of intense summer heat, or of long winters with only relatively short growing seasons.

The result of all these ecological factors which effects the density pattern of population, that is highly unbalanced, with such extremes as the Chengtu plain in Szechwan province, where the population density approaches over 1500 per square-kilometre, and vast areas of Sinkiang, Tsinghai and Tibet, which are virtually unpopulated. According to Leo A. Orleabs, the uneven distribution of the population is the fact that if a straight line is drawn from Ai-hur in Heilungchiang Province to Teng-Chung in Yunnan Province, only 40 percent of the land area of China lies southeast of this line, but 96 percent of the population lives there.

Volume of Overseas Chinese

The compilation of the total number of overseas Chinese is not an easy task. Since the countries take their official censuses at different times and intervals, it is extremely difficult to determine accurately the number of Chinese abroad in a particular year. Some countries use the criterion of ethnic background to define minority groups, others use nationally or citizenship.

Although there are Chinese in almost every country, they tend to be more numerous in the Tropical regions; in fact, about 97% of the overseas Chinese make their living in the areas between the Tropic of Cancer and the Tropic of Capricorn. The majority of overseas Chinese are found South of 18^0 N. The parallel that passes a few kilometers South of the island of Hainan. Besides the extremely dense Chinese population in South-east Asia, major concentrations occur in other tropical areas, notably in islands and countries bordering the Caribbean and on the islands of the Indian ocean and South Pacific. An overshelming majority of the Chinese abroad., about 96.6 percent are in Asian countries.

Table: *Distribution of overseas Chinese (various sources)*

Asia	***15,859,820***	***96.58%***
Thailand		3,799,000
Hongkong		3,197,081
Indonesia		2,545,000
Malaya		2,461,322
Singapore		1,302,500
Vietnam		1,035,000
Burma		450,000
Kampuchea		260,000
Sarawak		229,154
Macao		160,764
Philippines		151,759
N. Borneo		104,855
India		53,222
Japan		46,858
Laos		24,360
Korea		23,575

Brunei		21,795
Saudi Arabia		10,000
Timor		5,113
Turkey		3,300
Christmas Island		2,100
Pakistan		1,700
Ryukyu		785
Sri Lanka		450
Jordan		40
Afghanistan		28
Iran		14
Iraq		9
Lebanon		6
Oceania	*5,2572*	*0.32%*
Australia		19,800
New Zealand		9,500
Society Islands		6,948
New Gunia W.		5,000
Fiji Islands		4,943
New Guinea		3,000
New Britian		2,000
Nauru		800
Samos		301
Soloman Islands		200
Ocean Island		80
Africa	*43,735*	*0.27%*
Mauritius		23,266
Malagasy		8,901
South Africa		5,105
Reunion		3,000
Mozambique		1,735
Tanzania		522
Angola		500
Rhodesia & Nyasaland		303
Kenya		150
Uganda		70

Zanzibar		70
U.A.R.		27
Liberia		27
Congo		24
Ethiopia		18
Morocco		11
Cameroon		3
Nigeria		2
Europe	*20,586*	*0.13%*
Great Britain		12,000
Netherlands		2,400
France		2,000
Soviet Union		1,236
Denmark		900
Germany		800
Italy		313
Belgium		300
Spain		210
Portugal		143
Czechoslovakia		96
Poland		88
Austria		30
Sweden		24
Luxembourg		20
Greece		12
Switzerland		11
Norway		3
Latin America	*148,709*	*0.91%*
Cuba		31,039
Peru		30,000
Jamaica		18,655
Trinidad		12,000
Mexico		10,000
Brazil		6,748
Surinam		5,700
Guatemala		5,234

British Guiana		5,000
Equador		4,171
Costa Rica		3,000
Nicaragua		3,000
Panama		2960
Chile		2960
Venezuala		2580
Colombia		1400
Dominican Republic		1060
Honduras		817
Elsalvador		515
Aruba		486
Curacao		443
French Guiana		308
Argentina		240
Hati		204
Uruguay		152
Bolivia		35
Paraguay		12
Anglo-America	*295,489*	*1.79%*
United States		237,292
Canada		58,197

Will World provide any out lets? But the International Famine Relief Commission, after exploring various possible solutions, for the Chinese population problem, pointed out that "if other nations open wide their door the Chinese emigrants, if all the ships engaged in intercountinental passenger traffic on the seven seas were with—drawn from their usual routes and devoted themselves henceforth solely to transporting Chinese from their native land to other countries it is believed that they could not keep pace with the year-by-year increase of population."

Japan participated in the World War II not due to political problem but also due to the over population. Most of Japan is mountainous and agriculture is carried on only in 16% of the total land area of the country. The level and area is mostly confined to the Eastern coast of Japan. On the Eastern plains of Japan are located the great industrial centres such as Tokyo, Yokohama, Kobe-Oska-Kyato industrial regions. Here is found about one half of Japan's population. Japan's solving her increasing population problem by establishment of industries and trade. With the loss of Bangladesh Pakistan's population has been reduced from 182.42 million to 58.4 million and it has suffered a territorial loss of 1,43,327 km^2. As against 840,375 km^2, its area now is only 697,048 km^2. An interesting aspect of the population structure that has occurred as consequence of the emergence of Bangladesh is that there will be more Muslims in India than in Pakistan. Earlier out of 107 million total population it had a Muslim population of 93.7 million but now it will be only 42.8 million as against 60 million in India.

Sri Lanka's population is 16.3 million according to the latest Census, announced by the Department of Census and statistics, 1987. It represent 20.5 percent increase since 1963 when the last Census enumeration was done by way of sample survey. Districtwise Colombo has the heighest population with 7,699,392, representing a 22 percent increase in eight years followed by Kandy in Central Sri Lanka with 1,199,977, a percent increase. Citywise Colombo records the highest population with 563,705 followed by Dehiwala, a suburb of Colombo, with 154,313, and Jaffna, the capital of the predominantly Tamil-Speaking northern province, with 106,856. Five other cities have population ranging from 50,000 to 100,000.

In contrast to these densely populated countries, the South Eastern Asian countries such as Malaya, Burma, Indonesia etc. contain moderate densities per square kilometre. Burma is the most fortunate and thinly populated country of Asia. According to the census of 1987 the population of Burma was 38.8 million. The population pressure in south-east Asia is not so great on the cultivable land, excepting Java.

The island of Java is densely populated, and sometimes the density of population is over 500 persons per km^2. One reason for this dense population of Java is the fertile volcanic soils. Secondly, the Dutch colonizers made it their colony and developed plantation agriculture on the basis of local cheap labour. Although fragmentation is the Keynote of the topography of south-east Asia but demorgraphically a compact block of comparatively high density.

The western immigrants came and settled there permanently; crop yield per hectare was increased by the use of scientific techniques and agricultural implements. The climate is favourable for the crops especially for sugar cane cultivation throughout the year. Several other crops, along with rice, are grown on the same field.

If we compare the dense population of Java with the sparse population of Malaya. We find some cultural facts lying behind this unequal distribution. Plantation agriculture is the main factor to concentrated dense population in Java. But in Malaya the aim of westerners was not to settle there permanently but only to gain their economic end by the maximum exploitation of natural and human resources of that country. Other countries in the South-east Asia are still in various stages of economic development and political revolution.

In 1700, Russia's population of about 20 million was concentrated in the area that is approximately the central industrial regions of today. The subsequent dispersion occurred by migration to the South to the Volga area, and; ultimately, across Siberia. It is not by accident that in 1939 the masses of the Soviet Union's population lived in a broad triangle which had its base along the western border of the country and its apex in the Urals. A ecological map would reveal that this is the area comprising the steppe—which has fertile black-earth soils and, with certain exceptions, adequate rainfall—deciduous forests. Below this triangle in Central Asia, are the dry steppe and semi-deserts. Above it, in the north, are the coniferous forests and the frozen Tundra.

The direction and intensity of population movements in Russia were determined by the possibilities for agricultural development. Industrialization in a modern sense did not begin until the last quarter of the nineteenth century. It was then concentrated primarily in Central Russia, where manufacturing of consumers goods predominated, in the Baltic states and Poland, and in the Ukraine, where the abundance of Coal and Iron ore and a developed transportation net provided unusually strong incentives for industrial development.

Problem of over Population in Asia

Some countries of Asia are overpopulated. Infect there is no country which is over populated or underpopulated. It is national income that determines these relative terms. If the national income is greater and sufficient to feed the people of the country can never be called a overpopulated even that there may be 600 million people. But if the income is not sufficient to feed the countrymen even England, Japan and any other small island only for example, may be called overpopulated.

So far as Asia is concerned the overpopulation problem is not a new but for centuries back. And the overpopulation can be accounted due to three facts : Firstly, the insufficient national income due to unexploitation of natural resources :

Secondly, the agrigarian structure of population;

Thirdly, the immobility of the people.

Immobility of labour is due to two facts;

1. Love for their native land and kith and kins,
2. Lack of Territories.

Japanese attack on north China only to solve the problem of population but badly failed to get success. The overpopulation created many changes. It brings economic changes. When the people begin to starve they revolt against government and thus economic changes become the reason for political changes. Chiang Kai Shek was thrown out from the mainland of China only due to the economic crisis. Manchuria is still sparsely populated and has been a bone of contention among the nations for centuries.

Thus the overpopulation of Asia also results many crisis. And the solutions are the settle in the underpopulated area if (political ideologies permit) and development of industries, to explore the natural resources and as well as to develop of trade. Then and then only the population problem can be solved. There is also urgent need to check the growing population by family planning or some other preventive measures or used mechanical devices.

South-east Asia is the store of all sorts of natural resources. But no one had to utilize these resources. There had been lack of education, technicians and experts. This part of Asia called for labours.

'China a sleeping lion' according to Napolean, now awakened and they utilized vast natural resources of their country. The population problem as such was not a major point of discussion either by communist officials or by the country's press until 1954, when the results of the 1953 census registration were published. When the disclosure that the population of Mainland China was almost 600 million and with the release of sample vital rates which indicated a population growth of some 12 to 13 million annually, the demographic problem became a major concern.

Can China feed and generally take care of the millions? China's food production, taking into account some servere setbacks, grew faster than its population. China imports wheat but it also exports rice; the imports, moreover, have not increased over the last decade. Perhaps for these reasons and because of changing political attitudes, the theory of expansion for living-space is no longer "in" as far as with the standard of western countries are concerned.

Even when the theory was in fashion, the Chinese had embarked on a vigorous programme of family planning. The programme has had its ups and downs but it has been in operation for the greator part of the last 20 years. But the early 1960s, China appears to have succeeded in lowering its birth rate to about 2.5 percent. It also overlooked the fact that the countries around China's periphery are just as thickly populated as China proper. Moreover, almost two-third of China which lies outside the Great wall is virtually empty. Their region is admittedly difficult to open up for cultivation but not more so than, say. Soviet Siberia with the Chinese were supposed to cover for expansion.

The original Chinese target was to reduce population growth to one percent by 1970. That target has obviously not been achieved. At present, for the country as a whole, the growth may have come down about 2 percent. The current policy of energetically promoting late marriages, providing free contraceptives and facilities for abortion and sterlisation has, it is claimed already succeeded in urban areas. Peking is said to have brought down its growth rate of 1.24 percent. Shanghai has done even better with a record of only 0.69 percent. In rural areas, however, progress has been and will be slow but the trend towards decline in population growth is clear.

Of the three great nations of Asia, China has chosen the road of communism and a centrally planned and controlled economy; but Japan has chosen to build its future on a private enterprise system. Japan cannot feed itself, the cultivated area cannot be substantially increased: and the average farmer cannot live by farming alone. There is an estimated surplus far population of three of six million and a surplus of 1.3 million commercial and service workers. New entrants to the labour force, seeking employment for the first time, number 800,000 to 900,000 each year. Further industrialization is the hope for expanding the economy to absorb this surplus population, but raw-material resources are inadequate for the necessary expansion of industry. The two major factors that will determine Japan's ability to achieve full employment are capital investment and the export trade. Of the two, the increase of exports is the more difficult.

Population Distribution in Europe

According to Blache the European agglomeration of population begins where the Asian agglomeration ceases. It seems that more than two-third of the population of Europe forms a compact block of almost uniform high density.

The pattern of population distribution in Europe is mainly influenced by the ecological as well as cultural developments of that continent. There are human agglomerations in the river deltas but the agricultural population is not so dense as in India or China. Large human concentrations are to be found in industrial areas. The importance of coastal location for foreign trade is very remarkable.

The chief ecological factors controlling the distribution of population in Europe at present are climate, the relief and fertility of the land, and the distribution of mineral wealth, especially that of coal. But the cycle of population in any mineralized area is just as humanity increases in numbers by swarming, after the manner of bees, rather than by adhering in clumps like corals.....when the hive is too full, a swarm leaves it." Such is the history of population in mineralized parts of the earth of all time.

On the mainland of Europe, the coastal countries of English Channel and North sea the density of populations is high, which signifies the all round industrial progress of Europe. Here are found great Urban centres of France, Belgium and Germany; the valley of Rhine is also within these industrial Urban connections. The density of population in the coal fields areas of France and Germany is very high. According to Taylor, "The dense population in Europe is distributed along the coal belt from Swansea to Silesia. Here temperature is obviously not the main factor. Another obvious control (tending to prevent settlement) is elevation though here temperature is connected as well as absence of good soils and difficulty of communications. Hence, we shall find that the temperature effect is least complicated by other factors in the lowland regions where no coal is present." Belgium has along been plagued by the linguistic and cultural division between the Flemish-speaking population in the north and east and the French-speaking Walloons in the South. Until recently, economic and political control of the country has remained with the Walloons but the present increase of Flemish population seriously challenges this domination. A further threat has developed through the growing migration of industry from the South into the Flemish-speaking area. The gradual exhaustion of coal deposits in the Sambre-Meuse valley coupled with the discovery of important. New seams in the Kempenland has given impetus to this migration. Large tracts of Hitherto useless land have been reforested, and other areas, through the use of fertilizers and irrigation have been transformed into productive farm lands. Land reclamation is still going on in this crowded country, where the population density today exceeds 400 persons to a square kilometre.

Another population concentration is to be found cast of the Rhine valley along eastern Germany, and according to Taylor. In Europe the largest areas of heavy population occur in the Rhine and Po valleys below 150 m. and along the slopes of the Bohcmian Highland, etc., at about 450 m. only." In Spain and Bulgaria are there many inhabitants above 900 m. Holland, one of the most densely populated countries of Europe, has begun reclaiming land from the Zuider Zee. Poland has been one of the outstanding example in Eastern Europe of an agrarian economy unable to provide employment of many persons whose livelihood depends on agriculture.

The Balkan Peninsula is today one of the critical areas of the world. "As a direct result of world war II, there have been major changes in the political conditions, social and economic structure, boundaries, and territorial demands. This is especially true of Yugoslavia, on which world attention has been focused for some time. In 1946. Yugoslavia was reorganized into the Federal People's Republic of Yugoslavia. Postwar additions from Italy of Istria, Zara, the Cherso-Lussino island group, and the island of Lagosta increased the areas of Yugoslavia from 249,468 square kilometres to 256, 850 square kilometres. The population rose from 13,934,038 (1931) to 15,751,935 though there was an estimated direct loss of 1700,00 in world war II, comparatively one of the higher losses among the nations.

In Mediterranean Europe there is no extensive belt of human concentration like central Europe. It is due to the fact that the Mediterranean Europe is mostly mountainous. In Mediterranean countries, a great variety of crops are grown such as vines, fruits, early vegetables, tabacco, maize etc. Here the garden rather than the field was the focus of sedentary life. In the drier district of France, olives are grown. Here the main bases of population are agriculture and dairying.

In France the most thinly settled departments are Basses—Alps with 40 persons to the sq. km. and Hautes—Alps with 50, which again owe even these figures in part to their situation on the margin of the densely populated valley of the middle Rhone. "Manpower shortages already faced France before world war II, when almost 200,000 foreigners were being imported in some years. The same situation has now also become serious for other European countries like Britain, Belgium, Sweden, and west Germany (400,000 imported in 1990.) All of these countries have less than 20 percent of their population below the age of 15 years."

Norway, almost wholly a mountain country, average only 11 persons per sq. km. Less than a thousand sq. km. of its territory are under cultivation and these are distributed in small deltas at the heads of the fiords, in low strips here and there along its western coasts, or in the openings of its mountain valleys to the South-east. Here too is massed the larger part of its inhabitants. A barren granitic soil, unfavourable zonal location part of its inhabitants. A barren granitic soil, unfavourable zonal location, excessive rainfall, paucity of level land, leaving the "upright farms" predominant and remoteness from any thickly settled areas, together with the resulting enormous emigration, have combined to keep down Norways's population.

Finland, for the most part, is sparsely populated. It occupies a forested area where the chief means of livelihood is lumbering and most important motive power is horses. Hay is needed for these horses and milk and fresh meat should be produced for the increasing villages and forest population, many of whom are undernourished. There is little hope of growing grain economically so far north, but hay and dairy farming offers a real hope for raising the living standards. About 79% of the land to be cleared is tillable, and its nitrate content is uniformly high. The hay field are to be made large and free from obstructions so that farm machinery can be used everywhere.

In Europe the density of population depends on commerce, industries, trade and communication and mineral exploitation where as in Asia the concentration of population is based on agriculture.

In Great Britain, the sparsest population is found in the Sterile highland moors of Scotland, where the country of Southern upland has only 9 inhabitants to the square km. These figures reveal also the remoteness of a far northern location. In the Southern half of the island the sparsest populations are found in the Wales country of Radnor, Montgomery etc., with 30 to the sq. km and in English Westmoreland with 60, both of these mountain regions, but reflecting in their larger figures their close proximity to the teeming industrial centres of South Wales and Lancashire respectively.

The production of food differs in various parts of Britian according to the soil, climate and other ecological conditions. The population of Britain had become accustomed to a much higher standard of living, particularly to more meat and dairy products, which require more land for their production. Britain can supply only about half the population's food needs from its own farm land, it is important to get the best out of every hectare that is worth cultivating. The regions of densest population on the mainland of Europe are :

1. an area embracing the fertile basin of Rhine, Somme, Scheldt, and lower Seine, and possessing several coalfields;
2. an area stretching over much of central Germany, and extending South-east wards into Czechoslovakia and Poland, which is also rich in mineral;
3. the northern portion of Italy or Po valley, which is extremely fertile;
4. industrial attractions explain similarly high densities in North-western Europe.

The least populated regions are :

1. The northern portion of the continent, where the climate is cold and the soil relatiely unprodutie
2. South-eastern portion of Eurasia, which suffers from drought, and has a salt and sandy soil.
3. Alpineland and the Central plateau of Spain, where agriculture is impossible over large areas.

The whole of European Soviet Russia, as well as White Russia, the Ukraine, the Baltic states and the Balkan penisula have but a slight population compared with the countries of Central Europe.

Population of North America

In North America about 85% of the people are living east of 100° W. Longitude. The reason is that America was settled mostly by European immigrants and they could penerate the continent from the east. Hence human settlements firstly took place in the eastern Atlantic Seaboard. As the human pressure went on increasing these eastern human concentrations not only become dense but also spread westward and northward. In the west the density of population is higher only in particular areas where the irrigational facilities and mineral resources are available for human attraction.

Physiographically, Canada is a large country stretching from the Atlantic Ocean to an Pacific and from the United States to the Arctic, covering an area of 9,363,123 km^2. But it has a small population of 25,900,000. Canada is a land high mountains, vast forest (some of the world's largest), huge lakes, Solmon-filled rivers, kilometres of wheat lands and great mineral wealth, modern industries and breath-taking scenery. Consisting of extensive level plateau covering the area between the St. Lawrence river in the East and the Great Plains in the west and the Great Lakes in the South. The area gradually slopes toward the arctic ocean and extends all round the Hudson bay—called the

Candian or Laurentian Shield. It is one of the oldest parts of the earth's surface. Completely covered by layers of ice during the winter which melted and receded flooding the Hudson Bay and leaving a string of lakes. The largest is the Great Bear lake and Great Slave Lake where Eskimos and Indians still live by hunting and fishing.

The rocks here have vast mineral wealth and this area has gold, silver, and world's largest resources of uranium, nickel and asbestos. Timber is an important product from the Candian forests—amongst the world's largest. The timber is used for producing newsprint and other types of paper. The original inhabitants of Canada were the Eskimos and American Indian who form a fraction of the population which is mostly European (British) and French. There is also a sizable Indian community (mainly Sikh).

Population is densest in the food production area, and particularly in the interior continental plain including Prairies. The densest population in Canada are distributed along the food production belts. They are.

1. The Pacific area coastal of British Columbia and river valley;
2. Agricultural zone of Prairies provinces, and
3. Ontario peninsula and the St. Lawrence valley and the Eastern region drained by the St. Lawrence river as this is the industrial area where most of the people live.

Population is naturally exceedingly sparse in the cold regions of the far north, and in the marshy and moorland of Cent-ral Canada, as well as in those of North-West territories. It is fairly dense, and capable of considerable increase, around the sea-board of the Atlantic, in the Great lakes, and Western Pacific coast.

The United States has very much of a balanced economy which supports dense population. Over the whole of the United States of America the population per sq. km. averages 30, the mean density in the outlying territories being 400 and in the states 31. If we take the states as a unit and exclude cities, the mean density ranges between a minimum of 5 and a maximum of 160 per sq. km. There is a great variation in population density in the U.S.A. The Eastern half of the United States is more densely populated than the western. In the eastern half of the United States, the density populated areas are good agricultural regions such as the Mississippi valley, or those where commerce or industry or both, are highly developed. Areas with sparse populations in this part of the tract include through country such as parts of the Appalachians and poorly drained areas. The dense populations of the western half of the United States are in the valley of California and adjacent areas, the Willamette valley of Oregon, and the Puget Sound region of Washington. By contrast with those four states, Montana (4), Indaho (8) Wyoming (3), and Nevada, are sparsely inhabited. In these areas the land is too unproductive to support a flourishing agrucultural population.

In the continent of North America, Mexico is the only area that is least developed in spite of its varied potentialities and population is dense. The West Indies and Central America from the Tropical dependencies of the United States and Canada.

Distribution of Population in the Southern Continents

It is said that once upon a time there existed a big old continent known as the Gondwana land which later was broken into the present day three southern continents of South America, Africa and Australia. The three southern continents are economically backward, owing to backward and lack of communications. The southern continents are new continents with reference to the economic development and population problems. In the three southern continents the present day movements of population are from thinly peopled areas towards densely people areas rather than the reverse.

The variation in population density which is so apparent for the world as a whole, is similarly marked in South American countries. The present pattern of population distribution in the South American countries represents an evolution from that of the colonial period, when white inhabitants were few in number and settlement was confined largely to the Atlantic seaboard. This element of the interior and of the west coast has been accompanied by a westward shift of the centre of population. In the middle latitudes the density of population is due to the agricultural facilities such as are found in the plains of Agrentina, Uruguay, Paraguay etc. On the contray for British Guiana with about 2 persons per square km. its inhabitants concentrated on the lowlying coast lands and its interior largely devoid of people, remains very sparsely populated.

British Guiana, for the most part, is sparsely populated. Some 94% of the 377,000 inhabitants occupy only 4% of the area—the coastal strip. The remainder are scattered over the mountains, forests and Savannas that cover the rest of the country. Sugarcane is the main crop and has been the chief source of revenue since 1960s. Although there is little chance of further expansion because of the rationalization of the industry, present methods of production could be made more efficent. Rice is grown widely but for the most part is consumed locally. An expansion scheme is being attempted that would be needed to feed a larger population and for export to the West Indies. Among the mineral products are bauxite, second to sugar as on export, gold and diamonds. Forests cover some 1200 sq. kms and yield many types of timber of commercial value. Balata gum, used for insulation, and Coconut products are also obtainable. Improvement of diary and beef stocks could help the food supply, along with pork, poultry and citrus fruits.

A large population of the inhabitants find a livelihood in the grazing of cattle and the exploitation of the extensive forests. The greater part of west of South America consists of mountains and broken country where cultivation of an inferior kind is possible only in the narrow valleys and on the more gentle slops. It consists the hilly countries of Peru, Equador, Colombia, etc. One of its countries, Bolivia is the most sparsely populated. Venezula and Surinam in the north have also a very scantly population.

The greater part of Brazil is covered with forest and high hills. Although 71 percent of the population are agriculturists, they earn only 30% of the nation's total personal incomes, whereas 5% of the population receive 50% of the total. Brazil is a under populated country with promising frontier. The real Brazil is only to be found in the back country, in the thinly populated wildness beyond the frontiers of concentrated settlement called the "Sertoa". Although the population density is only about one third that of the United States. Throughout the remaining countries the conditions are fairly homogeneous, and the variations in the density are not very great.

Population of Africa

Although the continent occupies a fifth of the land area of the globe, it has only 7.5% of the population, 3 percent of the agricultural production, 9% of the metals, 5% of the rail-road and 5% of the trade of the world. The population distribution in Africa, is too marginal. Here density population is found along the Mediterranean coast the Nile Delta, in the Southern coastal regions, and Southern Zimbabwa. The total population of Africa at present is not known. This is because accurate enumerations are lacking for many areas.

The population of Ghana in 1974 was 9607000; in 1971 it was 9000,000 and in 1987 it was 13,700,000. Between the two censuses there was a net gain of 2.8% in population. This increase would have been made up of fertility rate, mortality rate and migration rate. But the rate increase was not the same for all parts. Western Africa includes Upper Volta, Togo, Sierra Leone, Senegal, Nigeria, Niger, Mali, Liberia, Ivory coast, Guinea, Ghana, Gambia and Dahomey etc. These countries form a distinct geographical unit, although in detail their geography must differ, they are in varying degree by distance from the Atlantic climatic influences. All these areas have high rural population densities to the square km. Ghana, Gambia, Nigeria have each more than 60 persons to per sq. km. These areas are depressed rural areas.

North Africa is regarded as the transitional climatic region between the Mediterranean north and the hot continental south, it contains the countries of the UAR. Tunisia, Sudan, Morocco, Libya and Algeria. The three—Morocco, Algeria and Tunisia—lie mainly among the Alpine mountain folds of Atlas; the other occupy a large part of the Great Nile plain, Libyan desert and the massifs of the central uplands with their interventing basins. Morocco was estimated to have a total population of 24,400,000 in 1987, thus ranking second in the Arab world after the UAR and seventh in Africa. Algeria is a country of marked physiographic contrasts, which set definite limits to its agriculturally productive potentialities. Where the conditions are naturally suitable for agriculture the soil is excellent. But considerable areas of the country are either desert or too arid for successful agriculture without being actual deserts in the usual sense of the term or salt marshes. The total population at the 1987 census was 23,500,000. Less extensive areas of high population density are found around coastal cities and small coastal plains. According to K. Sutton "Generally the coastal strip has above 50 inhabitants per sq. km. ranging up to a peak of just above 40,000 per sq. km. in the first and second arrondissements of the city of Algiers...More continuous extends of high density settlement are found in the Mitidja plain and Sahcl behind Algiers...Generally the communes of the high plains, steppes and Saharan Atlas show sparse population densities....Relative density as a population parameter loses its meaning in the Sahara departments, where high density oases contrast with almost total absence of population in vast areas, and seasonal nomadism adds a further difficulty to the study of distribution." The total geographical area of Egypt is 1,00,499 sq. km. with a population of 51900,000 (1987) and population density 36 persons per sq. km. The major portion of population resides in the villages and their economic prosperity depends largely upon the growth and development of agriculture. Density of population per habitable square kilometre has much more significance and according to Mountjoy it was as high as 843 in 1967.

The South African rural population is located mainly an arc that stretches from a point south of Bulawayo eastward through Shabani and Fort Victoria to Umtali and Salisbury. The higher densities in the native areas are in marked contrast with the neighbouring Europe farmlands. The arc is much wider to the north, in the upper Sabi valley, where rainfall is more reliable and more plentiful. West of the Bulaways-Salisbury axis the much smaller African population is more widely distributed, though there are local concentration at Dagamella, west of Que-Que, and in the Sipolilo Native Reserve, north west of Salisbury.

Sahara is the largest desert area of the world. Here the population is found only on the oasis where water is available.

Central and Middle is a region of mediocre fertility, without navigable waterways, large forests and even over immense stretches; a region broken up by deep ravines which, according to the season, are torrents of water or beaches of dry graval making it very difficult to establish good means of communication. On this rough, uneven soil the farmer has to fight not only against the difficulties that he meets in other climates but also against disastrous plagues which only too often come to destroy the fruits of the hardest soil and demolish the fairest hopes-factors like the irregularity of the climatic phenomena, exceptional dryness, torrential rains, in certain regions floods, grasshoppers, flies, epidemics and epizootics. Heavy concentrations of population are found in coastal areas, particularly in the drier, western part of the main islands of Pemba and Zanzibar. Here fragmentation of holdings, sever erosion, and declining crop yields, together with the non-availability of unused land of even moderate quality, all point to a situation in which continued population growth must give cause for grave concern.

Population Distribution in Australia

Geographically, Australia is an Indo-Pacific continent but demographically it is European. The discovery of Australia was delayed for a very long time because of its almost directly antipodal position of Spain, the great exploring nation of those times. The bulk of the present population of Australia consists of settlers from Europe and their descendants; but Chinese are found in some parts, notably Queensland, though their entrance is now rendered very difficult by various restrictions. The average annual rainfall over two-thirds of Australia's total area of 7 million square kilometre is less than 50 cms. Much of the Interior and western Australia is too dry for agricultural land and the humid tropical climate of the northern part has likewise been a deterrent to settlement.

The Northern territory, almost wholly a mountainous country and density is only 7 persons to the sq. km. Territories that have much agricultural land, but which have some industrial areas or are near large population centres, have a density varying from 600 to 1500 persons per sq. km. Included in this group are such territories as Tasmania, New South Wales and Victoria. The territories mainly dependent on agriculture only have under 300 persons per sq. km. An overwhelming proportion of Australian's population is today concentrated in the "crescent" of economic Australia.

The continent of Australia had large inhabited areas. In reality the population pattern in all the newly settled continents is everywhere same, because the coastal areas where the most suitable for the immigrants to settle. Later on with the increase in population, the inner areas were also gradually settled.; Australia provides a good example of this fact.

The temperate climate and the fertile soil, coupled with freedom from drought, make New Zeeland an ideal pastoral country with considerable population density. Appendix A shows the population data of the world.

5

Meaning, Nature and Scope of Population Geography

Meaning of Population Geography

Today, *overpopulation* is a word constantly on the tips of our tongues. It leaps forward as a ready explanation for many different irritating or frightening problems. You go to Yellowstone Park, and the ranger says you should have reserved your campsite last winter–overpopulation. You hear your neighbours arguing all night through the badly insulated walls of your apartment building–overpopulation. Population words bombard us constantly: the population crisis, the population bomb, the population explosion, zero population growth. People are compared to fast-breeding rats, the United Nations informs us that approximately four billion humans are alive on the earth. Overpopulation seems to hang over us all like a specter.

Would it surprise you, then, to learn that if people were evenly distributed over the earth's land surface, the resulting *population density* would be only about fifty-five people per square mile (twenty-one per square kilometer)? Of course, this is hardly the case, as all of you who live in less than one fifty-fifth of a square mile of personal spack know. In fact, anyone who has travelled from a large city to a small town realizes that humankind has distributed itself most unevenly over the earth's surface. Variations in population density range from zero in the uninhabited expanses of Antarctica to an average of 77,000 people living in each square mile of space (30,000 people per square kilometer) on the island of Manhattan.

Population geographers study these variations in density of people and the closely related "population explosion". So important is population distribution that some geographers regard it as "the essential geographical expression", a worthy focus for the entire academic discipline. Most cultural geographers feel that the uneven spatial distribution of people is a vitally important phenomenon, and it provides us an appro-priate point of departure for our study of cultural geography.

Nature and Scope of Population Geography

As a response to the stresses imposed on societies and their habitats throughout the world by unprecedented increases in human numbers and appetities in this century, there has been ever-growing interest by academics, general public and governments in 'the population question' at all levels of scale from global to local. Major international organizations like the United Nations, World Bank and Population Council have concerned themselves more and more with the study of global population problems, especially with a view to incorporating appropriate population policies in overall development strategies in the disadvantaged Thrid World. Meanwhile the governments of many Western nations feel they can no longer shelter behind their shield of affluence and ignore the population dimension. Several such governments, such as the United States, the Netherlands and the United Kingdom, have formed special advisory commissions or panels on population, although because population tends to be viewed in crisis terms rather than as a critical element of an ever-evolving social scene, these initiatives have been poorly funded, half-heartedly supported and sometimes aborted (Nam 1979).

By the late 1970s the more hysterical outbursts of concern about population growth had abated somewhat as the more gloomy scenarios for the future interaction of population and resources had been discredited by critical examination of their assumptions. Moreover, the growth rate in world population has fallen from about 2.0 percent per annum in the 1960s to 1.7 percent in 1980, essentially due to fertility falls in all developed countries and in some Third World countries, including most notably and dramatically the population giant of China. Yet such is the in-built momentum for absolute growth in population numbers globally and such are the uncertainties about the world's ability to provide supportive resources that population questions will remain a fundamental concern of mankind for decades to come. Even if developed countries myopically consider that their modest fertility levels cushion them from a survival crisis, they will certainly be concerned with a whole host of economic, social and physical planning problems stemming from the changing age structure of their populations. The study of human populations embraces a wide academic spectrum, and it is important to review the interests of at least some of the more central and important disciplines in the population field in order to assess how distinctive is the approach of population geography.

Demography

Within demography a distinction has commonly been made between formal or pure demography and population studies or social demography. The former emphasizes the demographer's prime concern with the collection, evaluation, adjustment, analysis and often projection of population data. A common core of analytical techniques comprises the standardization measures which control differences in population composition in order to pinpoint more precisely the nature of the dynamic processes. *Rigorous though* the methods of formal demography are, it has been suggested that 'technical pyrotechnics may displace penetrating analysis and that workers with high technical standards may become reluctant to work on scientifically important problems where precision is difficult or impossible to attain' (Hauser and Duncan 1959, p. 10).

Within the more general approach of population studies or social demography, the need to handle and analyse the basic demographic data objectively is not disputed, but this is regarded as the starting point in the exercise–a means to an end and certainly not an end in itself. Analysis proceeds to an attempted explanation of demographic patterns and processes from the subject matter and theories of social science disciplines like anthropology, economics, human geography and sociology. Indeed, the great majority of demographers have been trained in such disciplines before specializing in demographic research; few opportunities exist for undergraduate specialization in demography itself. The distinction then between a narrow and a broad approach to demography is a real one, although the labelling distinction is much less clear and, some would say, inappropriate. It would be difficult, for example, to make any distinction in content between the (two leading English language journals in the field, *Demography* and *Population Studies).*

Population Ecology

The study of human populations is regarded by some biologists as part of a single field of population study at all organic levels–microbe, animal, plant–which has been termed general demography, bio-demography or, increasingly, population ecology. Despite the diversity within plant and animal kingdoms, population ecology attempts to formulate generalizations, laws and processes for living organisms as a whole in their relations with each other and their habitats ecology' from the Greek valve meaning home on place to live.

This approach forms a useful context for an understanding of specifically human populations, particularly in the facilities it offers for field and laboratory experiments with animal numbers, not the point must be emphasized that man is a unique organism. Certainly man, like all organisms, functions in his environment by means of adaptation. He participates with other organisms in physical adaptation or natural selection. But man's brain size and associated mental powers have increased so much during the course of evolution that he has become uniquely able to adapt flexibly to his environment by modifying his behaviour, essentially through his cultural equipment. Culture can be regarded as learned behaviour that is socially transmitted: it includes customs, beliefs, technology and art. The culture gap between man and other animals has become so enormous that a quantitative difference along a continuum of organisms, can, in effect be regarded as a qualitative one.

Human Ecology

This emerged as a subject in the 1920s through the initiative of the Chicago school of sociologists, most characteristically expressed in the work of Park (1936). Much of their initial work was in applying the fundamental theories and methods of plant and animal ecology to the study of human communities, and, in particular, to interpreting spatial ordering within communities, especially metropolitan ones, through the process of competition or 'the struggle for space'. By the 1940s classical human ecology had become discredited as crude biological determinism, and the focus of the discipline swung more to studying the spatial distribution of social phenomena by a cartographic and statistical methods. Some sociologists would immediately use the term ecological to describe work which emphasizes a spatial perspective, particularly in the use of small area data (census tracts and enumeration districts). But more discerning commentators suggest that this limited approach, often benefit of theoritical justification, can hardly form the basis of an academic discipline. As Hawley (1944, p. 148) puts it clearly: 'Oue of the techniques employed in ecological research–mapping–has been mistaken for the discipline itself.

In more recent years human ecology has re-established itself by focusing on studies of the ecological complex at various community scale. From local global. The representation of the ecological complex of system which has perhaps achieved greatest impact has been that of. This indicates that human ecology focusses above all on the interactions between the fundamental dimensions in the complex from a functional point of view. Emphasis is given, therefore, to the way in which a population organizes and equips itself for survival in a particular habitat. Critically important to the human ecologist is an appreciation of the interdependence of the various dimensions; there is no unilateral causation, only mutual interaction. The lines in the diagram are thus meant to indicate linkages of functional interdependence. It follows that if dislocating forces affect any one dimension (e.g., concentrated mortality reduction, technological innovation or rapid environmental change), a whole chain reaction of changes is set up throughout the complex as a new equilibrium is sought.

Roots of Population Geography

It will be argued here that the nature of *contemporary* population geography owes more to the radical changes in approach and methods operating throughout human geography in the 1960s and 1970s than to the particular emergence and development in the 1950s of population geography as a recognized branch of geography; yet these early roots must at least be acknowledged and traced briefly.

In authoritative reviews of the evolution and nature of geography, Hartshorne (1939) and Wooldridge and East (1951) make not a single reference to population geography, thus indicating the late entry of population geography into the recognized systematic branches of the subject. Zelinsky (1966) attributes this to the late arrival and modest on the academic scene of demography (compared with, for example, the role of economics in stimulating economic geography and also to a dearth until fairly recently of population statistics, especially for subnational areas. Another factor would be the strong interests of many of the leading human geographers like Brunhes and Demangeon in settlement geography, where the distribution of population was studied through what was thought to be the more geographically acceptable medium of the cultural landscapel. Population *per se* was relegated to consideration in the more sterile forms of regional geography as part of a Place-Work-People chain, with a naive implicit assumption of an directional causation in that the physical environment was thought to strongly influence economic activities, which in turn controlled population patterns. It was in the early 1950s that population geography finally emerged as a systematic branch of geography in the sense that it dealt with a recognizably distinct group of phenomena and systematically related processes, the study of which involved a particular form of training. Much of the credit must go to Trewartha, who used the platform of his presidential address to the Association of American Geographers in 1953 to make a powerful case for, and an outline of, population geography (Trewartha 1953). He argued (p. 97) that population was the pivotal element in geography,m and the one around which all the others are oriented, and the one from which they all derive their meaning'. Other important formative statements were made at this time by P. George (1951) and James (1954a), and such was the growing interest in population geography that several synthesizing introductory texts appeared in the 1960s (P. Goerge 1959, Clarke 1965; Zelinsky 1966, Beaujeu-Garnier 1966. M.G. Wilson 1968, Trewartha 1969). Illustrative of growth in the subject at this time, papers in population geography increased from 3 percent in 1962 to 13 percent in 1972 of papers presented at the annual meetings of the Association geographical journals (Hansen and Kosinski 1973, p. 12).

Spatial distribution and areal differentiation of population attributes were clearly the unifying threads within population geography at this time. Thus Trewartha (1953, p. 87) saw its purpose as 'an understanding of the regional differences in the earth's covering of peoples' while James (1954a, p. 108) thought 'the objective of population geography is to define and to bring forth the significance of differences from place to place in the number and kind of human inhabitants'. The role of the population geographer was regarded by Zelinsky (1966, p. 5) as studying 'the spatial aspects of population in the context of the aggregate nature of places', and by Beaujeu-Garnier (1966, p. 3) as describing 'the demographic facts in their present-environmental context'. More explicitly, Clarke (1965, p. 12) state that population geography 'is concerned with demonstrating how spatial variations in the distribution, composition, migrations and growth of populations are related to spatial variations in the nature of places'.

In these traditional approaches a focus on spatial distribution was thought sufficient to distinguish population geography from demography, which is much more concerned with the intrinsic nature and universal attributes of populations and with a temporal, rather than spatial, dimension. But a problem had always existed in specifying population attributes appropriate for direct study by the population geographer. There is agreement on a core comprising distribution, density, ages sex and marital-composition, fertility, mortality and migration. A near rule-of-thumb way of circumscribing the field, suggested by Zelinsky (1966, p.7), is that attention should be confined to those human characteristics 'appearing in the census enumeration schedules and viral registration systems of the more statistically advanced nations' but inevitably there are data recording vantations between such nations, and, more critically, there is no theoretical justification for the suggestion.

Two questions have dominated the traditional approach of the geographer to population study: Where? and Why there? The first has been responsible for consideradle work in observing, identifying and, above all, depicting patterns of spatial distribution. Mapping of cross-sectional patterns thus dominated the early working population geography, well exemplified by James (1954a) devoting more than three quarters of his discussion of research fronties in population geography to problems of mapping. Various commissions of the Internatio-nal Geographical Union have been active in this area (William-Olsson 1963, Prothero 1972, Kormoss and Kosinski (1973) and many geographers have developed considerable expertise in population mapping, prompting one human ecologist, worried about the narrow focus of his discipline at one time on the mapping of social phenomena, to exclaim: 'It is sometimes difficult to understand why this kind of work should be called anything other than geography, except possibly–out of deference to the geographers–because of the inferior cartographic skill which is often exhibited' (Hawley 1944, p. 148).

Why there? takes the population geographer's approach a step further into an essentially ecological field, since 'the areal facts of population are so closely orchestrated with the totality of geographic reality' (Zelinsky 1966, p. 127). Consequently, the analysis and explanation of 'complex inter-relationships between physical and human environments on the one hand, and population on the other... is the real substance of population geography' (Clarke 1965, p.2). But for some time there has been a nagging unease among population geographers about what some regard (retrospectively, of course!) as inadequate theoretical and analytical rigour in much of their traditional work. The problem has been addressed explicitly by Woods (1979) it has attempt to introduce population geographers to a wide range of analytical methods from formal demography.

Modern Population Geography: its Changing Emphasis

The spatial and ecological approach to population phenomena remains the dominant and distinctive dimension of population geography, but the practice of the subject has changed appreciably in response to the winds of conceptual and quantitative change which have swept through human geography in the last two decades (M. Chisholm 1975, Johnston 1979). Nearly all of these changes were introduced and developed in branches of the subject outside population geography, most notably in economic and urban geography where the new outlooks and procedures were derived from other social science disciplines but their take-up by population geographers has been rapid and substantial–so much so that the methods they adopt have become less unique to demography and population geography, and more representative of the methods widely practised throughout human geography. Some might regret this as eroding the *raison, d' etre* of population geography, but an alternative view is that such integration of methods and communality of ideas augurs well for the health of the greater body, the parent discipline of geography. Indeed, this brings us back to an early statement by James (1954a, p. 108): 'To recognize population geography as a distinct topical speciality is not to think of it as separable from the whole field of geography'.

Several strands of methodological transformation can be indentified, although they are obviously more interactive and less compartmentalized than the following list suggests.

Quantification

The traditional geographical means of presenting and analysing population patterns has been essentially cartographic. Thus 'the map is the fundamental instrument of geographic research' (James 1954b, p. 9), with particular reliance on dot maps for population distribution and on choropleth maps for the distribution by area of population characteristics. The nature of such maps as essentially graphic devices greatly limits their use as analytical tools, and for explanation of patterns population geographers have often resorted to subjective and unreliable visual comparison of distributions in an attempt to assess correspondence and possible causal connection.

But with the growing awareness of quantitative methods within geography, an alternative means of describing numerical distribution within space has been adopted increasingly–the matrix–since this offers superior opportunities for subsequent analysis.

Each row of the matrix presents an inventory of a place or area, each column indicates how the incidence of a particular characteristic varies spatially, and each cell defines a geographic fact; thus cell ij in shows the value assumed by characteristic i at place j, and this may be expressed by the symbol x. Comparison of rows facilitates areal differentiation and therefore regional geography, while comparison of columns promotes the study of spatial association or covariation of phenomena, which is perhaps the most widespread general method currently adopted in population geography. The traditional visual comparison of map patterns has invariable been supplemented and often replaced by the use of correlation and regression methods on geographical data matrices (for a concise review of such methods and their applications, see D.M. Smith, 1975, chapter 8). Such methods allow spatial associations to be described with precision, and if analysis is extended into a consideration of residuals, hypotheses can often be generated about the identification and role of additional related variables. Nevertheless, one must be alert to a range of difficulties in applying standard statistical procedures like correlation and regression to spatially distributed data. There is the technical problem posed by the contiguity and spatial dependence of observations (Cliff and Ord 1973), and it has been shown that varying the size of the areal units can alter the value of the correlation coefficient (A.H. Robinson 1956, E. Thomas and Anderson 1965). More broadly, the distinction must always be made between the establishment of a measure of association and its use to infer a causal relationship; such a relationship must always be set within a plausible theoretical framework. In addition, there is the problem common to all forms of ecological correlation, that of making inferences about individual behaviour from essentially group or areal observations, as illustrated by the work of W.S. Robinson (1950) and Goodman (1959) on colour and illiteracy data in the United States.

The statistical analysis of areally based data sets is taken a step further into the realm of multiple correlation and multiple regression (Johnston 1978) by those population geographers who seek relationships between one dependent variable, whose spatial pattern they are seeking to explain, and several hypothesized independent variables. Additional examples of the advanced statistical methods increasingly adopted in population geography would be tedious at this stage when no reference is being made to specific findings.

Computergraphics

Increasing access to and familiarity with computers has stimulated the adoption, not simply of advanced statistical methods, but also of automated cartography. Population geographers have benefited from the use of computer mapping systems like SYMPA, which produces choropleth and isopleth maps by printing or over-printing characters or symbols to produce required densities. The advantages over conventional hand-drawn maps are essentially in accuracy and speed of production when maps of several variables required for a common areal base, thus allowing the initial programming and data preparation time to be effectively spread. Such mapping is particularly suitable, therefore, for the production of atlases portraying census variables; examples are provided by Rosing and Wood (1971) and Compton (1978a) in portrayals of census ward and enumeration district data for the West Midlands conurbation and Northern Ireland respectively. Automated cartography comes fully into its own where data are available for grid-squres. Such was the case in the 1971 UK Census when grid referencing of households was undertaken for the first time, thereby permitting the production of a population atlas based on square-kilometre data for Great Britain (Office of Population Censuses and Surveys 1981).

Models: Much of the traditional work in geography adopted a descriptive, historical approach to what were considered to be unique situations but growing familiarity with the methods of physical and social sciences has encouraged geographers to seek assiduously for patterns, regularities and order in space. This approach is expressed in the formulation and testing of generalizing concepts, law-statements and, above all, models. A model may be regarded as a representation of reality whose purpose is to give 'a conceptual prop to our understanding, and, as such, provide a simplified apparently rational picture for teaching and a source of working hypotheses for research' (Haggett 1972, p. 15).

The response of population geographers to this search for regularity was, initially, a qualified and hesistant one. Thus James (1954a, p. 112) described the important pioneering of Stewart (1947) on a formula to describe population grouping around urban centres in these terms: 'As a device for uncovering some kind of theoretical order in the processes leaking to the distribution of people this paper may be of great importance, although the geographer is more especially concerned with the modifications of the process in particular places.' Likewise, Zelinsky (1966, p. 24) warned that the complexity of population patterns 'denotes relatively strong individuality for specific areas and wide, if potenially explainable, departures from predicted patterns. There may, then, be no simple formulas to explain the geography of the world's population; at best only useful fragments of such laws may be available'. But more recent work in population geography, particularly in the study of migration as a form of spatial interaction, has conformed to the regularity-seeking trend in geography as a whole, as Zelinsky's own work (1971) on the mobility transition demonstrates.

Process Study: There has been a growing awareness in human geography that too much attention in teaching and research has been given to the observation and indentification of spatial patterns and too little attention to the processes which create and subsequently modify such patterns. More and more it is being appreciated that form and structure–the statics–are dependent on process and spatial interaction–the dynamics. Indeed 'in proper perspective, the distinctions we make between spatial process and spatial structure disappear because they are based upon a limited time perspective...Process and structure are, in essence, the same thing...When we distinguish spatial process from spatial structure we are merely recognizing a difference in relative rapidity of change...Properly considered, the spatial structure of a distribution is viewed as an index of the present state of an ongoing process' (Abler, Adams and Gould 1971, p.61).

Accordingly, geographers have become reluctant to inter process from structure, but ever more anxious to research the nature of the formative processes, not simply to explain past and present patterns but also to provide the basis of sound forecasting. Morever, an important logistical advantage noted by Brookfield (1973, p. 15) is that 'out fundamental shift away from the study of differentiation toward that of process carries with it liberation from any blanket-like constraint of scale in defining a geographical problem. The appropriate level of resolution becomes that at which the relevant process may best be recognized and analysed'.

Writing in 1967, Heenan (1967, p. 714) argued convincingly that 'population geographers have hitherto tended to concentrate their interest very heavily upon the end product or summation of change in preference to the study of those dynamic processes whereby change is wrought.' But more recently population geography has been among the most responsive branches of human geography to the new emphasis on process-oriented study. A collection of papers on population geography by Demko, Rose and Schnell (1970) is clearly designed to place reader attention on the process dimension, as are reviews by Clarkes (1973, 1977, 1978, 1979). The structure of this book has been designed specifically to recognize the prime explanatory importance in contemporary population geography of the dynamic components–fertility, mortality and migration–and the processes that fashion them;

it is their spatial and temporal interaction which produces changes in population numbers, distribution and composition. Interaction is all important, and none of the dynamic components and their associated processes can be studied effectively in isolation. Consider, for example, how appreciable emigration (selective in terms of yound male adults)from congested, famine-ridden Ireland in the nineteenth century promoted changes in the age, sex-and martial-structure of Ireland, which in turn had repercussion on the level of birth, death and marriage rates and on a wide range of social and economic structures.

It has been the spur of process study that has stimulated the theoretical, model-based and stimulation-oriented work on the spatial diffusion of ideas, behaviour and technology which arguably has been one of the very few significant strands of indigenous theory developed within geography as opposed to theory derived from other sciences. Diffusion studies of the type pioneered by Hagerstand (1968) have a clear relevance to an understanding of many of the key demographic process. It is possible, for example, to consider, although not necessarily to accept, that spatial patterns of both fertility and mortality reduction reflect the spatial diffusion of innovations like contraception knowledge and public health technology, with diffusion from metropolitan innovation centres being controlled by spatial and urban hierachical proximity.

Behavioural Geography: As a response to what some geographers regard as the excesses of macro-level deductive theorizing, of mechanistic quantification and of spatial fellshism, links are being forged increasinlgy with the behavioural sciences in order to appreciate better how the decisions which collectively from the basis of geographical processes and patterns are made. This approach has been used extensively and fruitfully by population geographers in studies of the socio-psychological determinants of migration–one of the most overt expressions of spatial behaviour. Wolpert (1965, 1967) provided the seminal geographical work in this field in his attempts to model the behavioural processes underlying the migration patterns which traditionally had been analysed by spatial scientists using little more than aggregate census data.

Applications and Relevance: Despite a widespread satisfaction in geography with the intellectual rigour imparted to the subject by recent conceptual and technical developments, there exists an equally widespread concern that the subject has hot been contributing powerfully to solving the pressing problems facing mankind. House (1973, p. 273) maintains that 'particularly disadvantageous to geography is the widespread inability or disinclination by administrators to view problems, decisions or policies in a spatial framework, or even as having an important spatial implication'. Yet others, like D.M. Smith (1977, p. 368), feel that 'the spatial or areal perspective imbued by geographical training may itself impose blinkers that obscure the operation of 'non-geographical" variables'.

Population geographers in the Soviet Union have been centrally involved in settlement and economic planning (Clarke 1973), but in the West their applied role has been more tentative (Udo 1976). Yet a modest growth in applied and socially relevant work can be illustrated by reference to three areas. First, effective development planning in the Third World is enhanced by a full appreciation of the spatial population patterns traditionally prepared by population geographers from census and other data; examples from Africa include the work of Hilton (1960), Clarke (1966), I.D. Thomas (1968, 1972) and Gould (1973). Second the geographer's interest in optimization or 'best location' has important applications in the population field. Work by Robertson (1972, 1974) is an example of the application of statistical and cartographic methods to a matrix of population data on a regular grid to give precise yet flexible answers to problems of facility location when maximum accessibility of population to facilities is desired. Third, significant contributions are being made by some population geographers to the construction of spatially disaggregated population forecasting models (Rees and Wilson 1977).

Ideology

Often associated with calls for a more socially relevant, humane and welfare oriented approach in geography have been structures about the value judgements implicitly adopted in academic studies. Harvey (1974) has demonstrated powerfully, with particular relevance to the then active debate on the global population resource balance, that in geography as elsewhere there is no such thing as an ethically neutral scientific methodology, and Hurst (1973) has accused human geography of largely supporting the status quo and predisposing people to an acceptance of the social ills accompanying capitalism.

Perhaps the area of population geography most influenced by a modern radical perspective has been that of development-related population processes in the Third World. Brookfield (1973, 1975) has shown that geographers, because of their deep commitment to Eurocentric values, have uncritically viewed development as a desirable process of diffusion of Western culture and technology to less favoured areas. It is now being recognized that alternative explanatory models are available, notably the inequality-perpetuating 'core-periphery' or dominance-dependency' relationship, and that goals like the achievement of self-respect and self-reliance might be considered no less desirable in a meaningful development process than material gain. One radical geographer has also argued with reference to the Third World that 'the poverty which is regarded as symptomatic of reckless population growth is rather a *structural poverty* caused by the irresponsible squandering of world resources by a small handful of nations' (Buchanan 1973, p.9) and that the current Western promotion of birth control programmes arises less from genuine humanitarianism than from the attempt of white northern imperalism to preserve the global status quo and its own privileged access to the resources of the dependent international periphery. Whether or not a Marxist mode of analysis gives more perceptive insights to population study than those grounded in capitalism is perhaps a less important matter to population analysts than the need to at least recognize and acknowledge the guiding framework of one's ideology and values.

6

Population Change

Changing Distribution of World Population

Population growth will be concentrated in certain regions; elsewhere, human numbers will stablize or even decline. Within countries, populations will continue to shift from rural to urban areas, while becoming increasingly older and better educated. Migration between countries will be an increasingly important factor in international relations and the composition of national populations.

Regional Distribution Changing

As the global population has doubled over the past 40 years, the shifts in geographical distribution of that population have been equally remarkable. In 1960, 2.1 billion of the world's 3 billion people lived in less-developed regions (70 percent of the global population). By late 1999, the less-developed regions had grown to 4.8 billion (80 percent); 98 percent of the projected growth of the world population by 2025 will occur in these regions.

Africa, with an average fertility rate exceeding five children per woman during the entire period, has grown the fastest among regions. There are almost three times as many Africans alive today (767 million) as there were in 1960. Asia, by far the most populous region, has more than doubled in size (to over 3.6 billion), as has Latin America and the Caribbean. In contrast, the population of Northern America has grown by only 50 percent, and Europe's has increased by only 20 percent and is now roughly stable.

Africa's share of global population is projected to rise to 20 percent in 2050 (from only 9 percent in 1960), while Europe's share is projected to decline from 20 to 7 percent over that same period. In 1960 Africa had less than half the population of Europe; in 2050 it may be approaching three times as many people.

The altered balance of population distribution among regions does not in itself pose a problem, so long as development progresses everywhere and population growth is balanced by the development of social and economic capacity. The challenge remains to create conditions that will enable countries in all regions to adopt policies and strategies that foster equitable development.

Global Trend Towards Urbanization

The movement of people towards cities has accelerated in the past 40 years, particularly in the less-developed regions, and the share of the global population living in urban areas has increased from one third in 1960 to 47 percent (2.8 billion people) in 1999. The world's urban population is now growing by 60 million persons per year, about three times the increase in the rural population. Increasing urbanization results about equally from births in urban areas and from the continued movement of people from the rural surround. These forces are also feeding the sprawl of urban areas as formerly rural peri-urban settlements become incorporated into nearby cities and as secondary cities, linked by commerce to larger urban centres, grow larger. The proportion of people in developing countries who live in cities has almost doubled since 1960 (from less than 22 percent to more than 40 percent), while in more-developed regions the urban share has grown from 61 percent to 76 percent. There is a significant association between this population movement from rural to urban areas and declines in average family size.

Asia and Africa remain the least urbanized of the developing regions (less than 38 percent each). Latin America and the Caribbean is more than 75 percent urban, a level almost equal to those in Europe, Northern America and Japan (all are between 75 and 79 percent).

Urbanization is projected to continue well into the next century. By 2030, it is expected that nearly 5 billion (61 percent) of the world's 8.1 billion people will live in cities. The less developed regions will be more than 57 percent urban. Latin America and the Caribbean will actually have a greater percentage of inhabitants living in cities than Europe will.

Cities and towns have become the engines of social change in all regions. Their rapid growth presents opportunities for future development but also serious challenges. Urban population growth has outpaced the development of employment, housing, services and the rest of the social and physical infrastructure. Poverty persists in urban and peri-urban areas, suggesting a failure of policies to ensure an equitable distribution of the fruits of development. Numbers of poor women, in particular, have increased, both in urban areas (where work opportunities remain limited) and in rural areas (where women are increasing being left behind by husbands or children seeking urban opportunities).

As people have moved towards and into cities, information has flowed outward. Better communication and transportation now link urban and rural areas both economically and socially. The result is that the ecological and sociological "footprint" of cities has spread over ever-wide areas, creating an urban-rural continuum of communities that share some aspects of each lifestyle. Fewer and fewer places on the planet are unaffected by the dynamics of cities.

The spread of mass media has also blurred the rural-urban divide. New ideas, points of reference, and life possibilities are becoming more widely recognized, appreciated and sought. This phenomenon has affected health care, including reproductive health, in many ways. For instance, radio and television programmes that discuss gender equity, family size preference and family planning options are now reaching formerly isolated rural populations. This can create demand for services, higher contraceptive use, and fewer unwanted pregnancies.

Globally, the number of cities with 10 million or more inhabitants is increasing rapidly, and most of these new "megacities" are in the less-developed regions. In 1960, only New York and Tokyo had more than 10 million people. By 1999, the number of 17, 13 in less-developed regions. It is projected that there will be 26 megacities by 2015, 22 in less-developed regions (18 will be in Asia); more than 10 percent of the world's population will live in these cities, up from just 1.7 percent in megacities in 1950.

Components of Population Change

Patterns of distribution and density represent only a starting point in the geographical study of population. The dynamic aspects of the study, the changes in distribution patterns and the interaction of the demographic processes of births, deaths and migration, also form an important field of geographical study. A discussion of these processes, referred to collectively as the components of population change, forms the subject of the present chapter.

Fertility

The term fertility refers to the occurrence of live births among a defined population. In most parts of the world fertility exceeds both mortality and migration, and is thus the main determinant of population growth. At the same time, fertility is more difficult to analyse than mortality and is subject to greater short-term fluctuations. Whereas mortality is inevitable and involuntary, fertility can be controlled and is determined by a wide range of social, economic and political influences as well as physiological and psychological factors.

Various indices are employed to express fertility. The simplest, but in many ways the least satisfactory, is crude birth rate. This is simply the ratio between number of births (usually in one year) and total population (usually a mid-year estimate for the year in question). Thus the crude birth rate for the UK in 1986 was

$$\frac{755,000(\text{total number of births})}{56,763,000(1986\text{ mid-year estimate of population})} \times 1,000 = 13.3\%.$$

Crude birth rate has the advantage of being easy to calculate and involves data that are generally readily available, if not always highly accurate. However, it fails to take into account the age and sex composition of the population, and this reduces its value for purposes of comparison. To overcome this deficiency other measures have been devised.

A standardised birth rate involves the calculation of what the birth rate for a region would have been if its age composition had been the same as that of the country as a whole. The computation is tedious but produces a rate which is directly comparable with that of other regions since variations in the number of births resulting from differences in age structure have been eliminated. Another useful index is the general fertility rate, which also avoids some of the deficiencies of the crude birth rate by changing the denominator from total population to the number of women in the reproductive age group, usually defined as 15-45 or 15-49.

The World Pattern of Fertility

In 1985 crude birth rate for various nations of the world ranged from 9.6% (West Germany) to 55.1% (Kenya). The highest levels of fertility, with crude birth rates in excess of America, of 40%, are experienced in most parts of Latin Africa, the Indian subcontinent and South-East Asia. Very few of the developing countries have achieved any success in making the transition from high to low fertility. On the other hand, relatively low fertility, with crude birth rates of less than 20%, is typical of the developed countries of Europe, North America, Oceania, USSR and Japan. These facts have led certain writers to suggest that a decline in fertility is an inevitable corollary of economic and social advancement.

Among those countries experiencing low fertility, the reduction of former high birth rates has generally been achieved since the late nineteenth century. An interesting exception is Japan, which more than halved its birth rate in 15 years from 35% in 1947 to 17% in 1962.

Detailed study of the population in those regions which have reduced their crude birth rate figures shows that the apparent decline in fertility is in fact a real one, and not simply a reflection of changes in population structure such as a reduction in the number of women of child-bearing age. Furthermore, although there is not complete unanimity on the point, there is nevertheless general agreement that there has been no recent reduction in the fecundity of populations in the world's economically advanced nations. In other words, we are not considering an involuntary decline in fertility but rather a situation in which a voluntary or controlled birth rate may be said to exist.

Factors Influencing Fertility Levels

Even in those parts of the world where the majority of the population has passed from what may be termed a 'natural' birth rate to a 'controlled' birth rate, human choice is not entirely free. Decisions concerned with family limitation are both consciously and subconsciously influenced by a wide range of moral, intellectual, financial and social motives. The changing status of women in society, new attitudes towards children and marriage, the decline of religious beliefs and superstitions in many parts of the Western World, and the prevalence of material ambition in modern society must all influence attitudes towards family size and thus indirectly affect birth rates.

Certain influences on birth control may be considered more closely. The factor of religion is important. Most of the world's major religions encourage family development and are opposed to birth control, sterilisation and abortion. Strongly religious communities thus tend to have high birth rates. Muslims, for example, especially those living in economically backward countries with a traditional society in which religious dogma remains unquestioned, are generally characterised by high birth rates. Libya and Saudi Arabia, both predominantly Muslim, had birth rates of 46% and 43% in 1985. In Western Societies Catholic communities generally have higher birth rates than other religious groups. In Quebec, with its Catholic French Canadian population, the crude birth rate was one third higher than that of neighbouring Ontario, which is largely Protestant, during the period 1920 to 1950.

Correlations have been shown to exist in many countries between the duration and level of education and family size. It has been shown in a number of studies that the more advanced the level of educational attainment of parents the smaller the number of children per family. Such a situation is probably related to the negative correlation that has also been demonstrated between family income and the average number of children per family, especially among the middle classes. It has been said that limitation of the family is at its strictest where the difference between the actual and desired standard of life is greatest. The poorest people, possessing little, also have limited ambitions, and are typified by high birth rates. In contrast the middle classes, although having only modest means, have the greatest material aspirations and are frequently motivated by the social pressures which characterise modern advanced societies. Thus, it is this social class which generally has the lowest levels of fertility. In contrast, the well-to-do with large incomes and enjoying a social status which they can assure for their children generally have relatively high birth rates.

At national level, cycles of economic prosperity and depression appear to have a marked influence on marriage and birth rates. Correlations are not precise but in general terms it would seem that a sudden onset of prosperity favours an increase in family size, while a gradual rise in living standards has the reverse effect. The reasons for this are not altogether clear, but it may be that in the latter situation families become accustomed to improved living standards and do not wish to involve themselves with the expense and sub-division of income which children necessitate. On the other hand, economic depression, with accompanying unemployment, causes a sharp reduction in birth rates. These relationships are well illustrated by reference to statistics for the USA during the period 1920 to 1970.

During the present century political influences on population growth have emerged as a factor of some importance. Numerous examples exist of governments trying to influence their pattern of national population growth for economic and strategic reasons. During the late 1930s Germany, Italy and Japan all tried to increase their rates of population growth by offering financial inducements and concessions to those with large families; Australia has actively sought to boost its population by financing selective immigration. Conversely, the governments of India and Pakistan are promoting and encouraging programmes of birth control in an attempt to check their high rates of population increase, which have long counteracted economic progress.

It will be appreciated that these influences operate chiefly in societies where there is a high level of appreciation of economic change with appropriate responses and adjustments at the individual level. What then are the influences on birth rates in those parts of the world which have not made the transition from 'natural' to 'controlled' birth rates? What factors determine fertility levels in those parts of the world with crude birth rates in excess of 40%?

It is unfortunate that the nations of the world that experience the highest birth rates generally have the least complete and reliable statistical information. Nevertheless, there is sufficient evidence to suggest that social customs and taboos rather than physiological or environmental factors are the chief determinates. Concepts of marriage, monogamy or polygamy, variations in the average age of marriage, as well as religious proscriptions, all have an important bearing upon fertility levels. For example, in the Hindu society of India, girls marry on average at 16 years of age and produce their first child at 18. The Hindu religion places special importance on a male heir in a family, with the result that a large family may be produced before this is achieved. Furthermore, in order to ensure the survival of a male heir under conditions of high infant mortality, the family may be further increased. Thus, in many parts of India there is an average of seven children per family. In other words, reproductive behaviour is largely conditioned by social and religious customs which favour high fertility.

With the exception of the feeble reproductive capacity of people living in high-latitude regions, there is no evidence to suggest that climate or other environmental factors have any direct bearing upon fertility. Numerous localised studies of birth rates have revealed racial differentials, but there is no consistent pattern. For example, in Brazil the White population has a higher reproduction rate than the Negro population, while in the USA the reverse is true. It seems, therefore, that where such differentials exist they are probably a reflection of differences in social class and economic status, the effects of which have been noted earlier.

Relationships between fertility and diet and health have been examined by Josue de Castro in his book, *The Geography of Hunger*. He has drawn attention to the fact that the poorest and most undernourished people are usually characterised by the highest birth rates. This rapid increase in numbers further accentuates the poverty, thus creating a vicious circle which is extremely difficult to break. At the same time, of course, these underprivileged and impoverished people also suffer from the highest rates of infant mortality and lowest expectation of life, which partially counteract the results of high fertility rates.

Another factor which directly affects birth rates is population structure, especially the age composition of a population. Comparison of birth rates and fertility rates will reveal the importance of this factor. For example, in 1982 Singapore and Spain had similar crude birth rates (17.2% and 15.2% respectively), but Singapore's fertility rate was much lower (58%) than Spain's (73.1%) because her age composition was more youthful. In other words, areas with a high proportion of young adults may be expected to have high birth-rate figures. New towns, pioneer settlements and regions with high immigration rates tend to fall into this category. For the same reason urban areas often have higher birth rates than their rural neighbours, although their fertility rates may not be dissimilar.

We may conclude that the factors influencing the level of fertility in any area are largely economic, social and cultural rather than physical. Analysis is made difficult by the complex inter-relationships that exist between the factors concerned, and also, at a purely practical level, by the deficiencies and inaccuracies of much of the available data. It is fortunate that mortality statistics are generally more continuous and reliable than those relating to fertility, although again there are many uncertainties and problems.

Mortality

The term mortality is used to describe the occurrence of deaths among a defined population. Infant mortality is generally defined as the number of deaths of infants under the age of one year expressed per thousand live births. It has been described as 'perhaps the most sensitive barometer of the fitness of the social environment for human life' (Lewis Mumford). As a result of medical progress and improved health services, most nations have been characterised by declining mortality rates during the present century, a fact which contributes largely to the so-called 'population explosion' of modern times. The geographical study of mortality has attracted much attention since 1950 and medical geography, which concerns itself with the effects of environment on spatial variations in mortality, has emerged as an important field of specialised study.

As with fertility, a number of indices are employed to express levels of mortality. Crude mortality rate is simply a ratio between numbers of deaths and total population. Its manner of calculation may be demonstrated by the 1986 figure for the UK

$$\frac{661{,}000 \text{(total number of deaths)}}{56{,}763{,}000 \text{(1986 mid-year estimate of population)}} \times 1{,}000 = 11.6\%.$$

Crude mortality rates are greatly affected by the age structure of a population and a standardised mortality rate, calculated in the same manner as the standardised fertility rate referred to earlier, is essential for comparisons of mortality.

Life tables, which were first compiled for insurance purposes and are concerned with the effects of present mortality rates on the future age and sex composition of population, are also widely employed in the analysis of mortality. They involve following a hypothetical generation of births throughout life, subjecting it to varying mortality rates at different ages and plotting the number of survivors at each age. The life table mortality rate, which may be calculated from such a table, represents the probability of dying at a given age. It decreases to a minimum value at about the age of ten years and then increases steadily with advancing age. Another measure which may be extracted from life tables is that of expectation of life or average length of life. Expectation of life at birth is a particularly useful index since by its nature it is a standardised measure, unaffected by the age structure of the population in question.

The World Pattern of Mortality

Crude mortality rates in 1985 ranged from 2.8% (Kuwait) to 29.7% (Sierra Leone), with a mean value of 14%. Countries with mortality substantially above the world average are located chiefly in South-East Asia, Africa and parts of Latin America. Populations in these parts of the world are generally characterised by high infant mortality (up to one-third of all deaths may occur in the first year of life) and low expectation of life resulting from poor standards of diet, housing, hygiene and medical care. Conversely, low mortality rates are usually related to high living standards, good medical services, youthful population structure or a combination of these factors. It is interesting to note that virtually the same nations experience both the lowest mortality rates and the lowest fertility rates. Crude mortality rates of about 10% are typical of most countries of Western Europe, North America and Oceania, together with the USSR, Japan and certain countries of South America (Argentina, Uruguay and Venezuela). A number of small territories have also succeeded in reducing their mortality to surprisingly low levels; these include the Bahamas (5.0%), Fiji (4.6%), Singapore (5.2%), and Hong Kong (4.8%).

A number of points should be emphasised at this stage. On a world scale the crude mortality rate is only about 40 percent of the crude birth rate. In many countries the great gap that exists between fertility and mortality is producing an extremely rapid natural increase of population and placing a severe strain upon natural resources. This problem will be examined in Chapter Eight. At the same time world variations in mortality are smaller than variations in fertility. Among the economically advanced nations mortality decline has now been checked by the ageing of population caused by low fertility, while in many developing countries there have been spectacular reductions in mortality rates resulting from improved medical services, campaigns against infectious diseases and in some instances by economic progress and the raising of living standards. It has been said with some truth that there is now more equality in the face of death. If these trends continue, the influence of mortality on population growth will be further reduced and fertility will become even more firmly the main determinant of population growth.

Differential Mortality

The same basic influences on mortality which operate on a world scale, such as variations in standards of nutrition, medical care, hygiene and housing, also act as differentials within particular nations. In many instances they operate through differences in Social class; that is to say, the underprivileged sections of a population are obliged to reside in substandard or slum housing and may be unable to afford an adequate, balanced diet, proper medical care and hospital treatment. Thus, the Black population of the USA has a substantially higher mortality rate at all ages than the White population, while in South Africa the level of infant mortality of the Black population is almost six times higher than that of the White population. In those countries which have a well-developed system of social security and welfare the class differential influences mortality less strongly. Workers enjoy the benefits of the same medical services as more fortunate individuals and may be housed in new towns or modern residential developments. Such concepts are fundamental in the planning of socialist and communist regimes.

Even in societies lacking extreme social differentials there are usually considerable internal variations in mortality. Although statistics are difficult to obtain, a number of studies have demonstrated the importance of occupation structure as an influence on mortality rates. It is obvious that certain occupations involve far greater hazards than others. Large mortality differences between husbands and wives will confirm that the essential influence is occupational rather than the product of living standards. Causes of death may also be indicative. Miners, for example, are vulnerable to a high level of accident risk as well as being prone to respiratory diseases such as pneumoconiosis and tuberculosis. Since many occupational groups tend to form residential concentrations it follows that occupation structure can constitute an important localised influence on mortality patterns.

Mortality also varies according to place of residence. Even after adjustments have been made for differences in age and sex structure, most urban areas still show higher mortality rates that rural areas despite their superior provision of medical services. This is especially true of large cities and conurbations. The reasons are not difficult to find; high population densities, crowded living conditions, high traffic densities, atmospheric pollution and the nervous strain of urban living all contribute to relatively high mortality rates. Mortality rates in the highly urbanised and industrialised regions of the UK, for example, are significantly higher than those in the rural areas.

Causes of Death

It is surprising to learn that even at the present time many countries make no attempt to compile detailed records of causes of death. In fact, countries that collect and publish such data represent less than half the total world population. Even the available data must be treated with great caution. Problems include vague declarations of cause of death such as 'senility', inaccurate diagnoses such as the presentation of a single cause when multiple causes operate, lack of uniformity in disease nomenclature which renders international comparisons difficult, and the fact that many causes of death have only recently been recognised, which invalidates historical comparisons. The World Health Organisation has sought to reduce these problems by the publication of its *Manual of the International Statistical Classification of Diseases, Injuries and Causes of Death* and by encouraging the compilation of accurate mortality statistics, but much remains to be done in this field of data collection.

Causes of death may be divided into two categories: exogenetic and endogenetic. The former are essentially the result of environmental influences and include infectious, pulmonary and digestive diseases associated with inadequate and contaminated food and water supply, and low standards of housing and hygiene. The latter, which are sometimes referred to as degenerative causes, are essentially biological and include congenital diseases and the gradual exhaustion of body functions.

***Table:** UK: Principal Natural Causes of Death (expressed per 1000 deaths)*

Average	***1848-72***		***1986***
Infectious diseases	321	Diseases of the circulatory system	480
Tuberculosis	146		
Scarlet fever	57	Heart diseases	358
Typhoid	38	Cerebrovascular diseases	122
Respiratory diseases	148		
Bronchitis	66	Cancer	242
Pneumonia	57	Respiratory diseases	108
Diseases of the nervous system	129	Pneumonia	48
		Bronchitis	21
Diseases of the digestive system	83	Diseases of the digestive system	31
Diseases of the circulatory system	53	Diseases of the nervous system	19
Other causes	266	Infectious diseases	4
		Other causes	116

A typical feature of demographic evolution is for a country to pass from a situation in which exogenetic causes of death predominate to one in which endogenetic causes are dominant; that is to say, where disease is little controlled, environmental causes hold sway, but with improved medical services these are gradually eliminated, leaving heart diseases and cancer as the main killers. Such an evolution may be demonstrated by reference to statistics for the UK.

In the mid-nineteenth century one person in three died from infectious diseases; in 1986 only one in 250. On the other hand, among the endogenetic causes, deaths from failure of the circulatory system, notably heart diseases, together with cancer have risen dramatically in recent decades and together accounted for over 70 percent of deaths in 1986. A similar evolutionary process could be presented for most countries with mortality rates of less than 10%. In most cases the changes have taken place over more than a century, but in a few instances the evolutionary sequence has been completed far more rapidly. Causes of death in Japan, for example, showed a complete reversal in the proportions of exogenetic and endogenetic causes between 1945 and 1970.

The struggle against death, which has occupied man since earliest times, appears to have entered a new phase during the present century. In most parts of the world there has been a definite reduction in mortality rates, usually accompanied by changes in the pattern of causes of death. The consequences of this development are clear: longer expectation of life and reduced infant mortality. Both have been hailed as successes, but at the same time they present new problems; namely, an increase in the number of elderly people and more mouths to feed. Unless success in the reduction of mortality rates is matched by resource development it must inevitably accentuate existing poverty and suffering.

Migration

Migration, the third component of population change, may be interpreted as a spontaneous effort to achieve a better balance between population and resources. Migration is defined here as a movement of population involving a change of permanent residence of substantial duration. Thus, no reference is made at this point to nomadism, transhumance, tourism, commuting or other similar movements which lie outside the scope of such a definition. Although there is an enormous diversity of types of migration in respect of cause, distance, duration, volume, direction and organisation, it will suffice for present purposes to draw a simple distinction between international migration and internal migration, the latter involving no crossing of international boundaries. A basis problem in the enumeration of migration relates to the distinction between short-term migrants and temporary visitors. How long must a visitor reside in a country before he is classified as an immigrant? Indeed, migration statistics must be treated with great caution, for definitions and methods of data collection vary widely from country to country.

From the evidence of those countries with reliable statistics it is clear that migration is generally a selective process; that is to say, there is a tendency for certain sections of the population to be more involved in migration than others, although this selectivity is less marked than formerly. Unless special circumstances operate, the majority of migrants tend to be young adults; international and long-distance internal migrants are predominantly male, while females are more involved in short-distance movements; single persons and married couples without children move more frequently than couples with children; professional workers are more mobile than unskilled workers.

Internal Migration

Very few countries have satisfactory data on movements of population within their boundaries. In Norway each commune has a register showing the destination of all those leaving the district and the area of origin of all new-comers, but such direct enumeration is extremely rare. More typically, internal movements of population must be inferred by indirect methods. For example, by comparing successive census totals it is possible to measure net migration balance by subtracting natural increase from total intercensal change. Alternatively, comparison of place-of-birth statistics with present residence may be suggestive of patterns of movement. However, such methods provide only a very incomplete and inadequate picture and in many instances geographers are obliged to conduct their own sample surveys.

Much attention has been directed towards the formulation of theories to explain patterns of migration. As early as the 1880s E. G. Ravenstein made a detailed study of migration statistics and presented papers on 'The Law of Migration' in which the following principles were proposed (presented here in summary form):

1. The great body of migrants only proceed a short distance and consequently there takes place a universal shifting or displacement of the population, which produces 'currents of migration' setting in the direction of the great centres of commerce and industry which absorb the migrants.
2. Migration occurs in a series of stages. The inhabitants of a country immediately surrounding a town of rapid growth flock into it; the gaps thus left by the rural population are filled up by the migrants from more remote districts, until the attractive force of a rapidly growing city makes its influence felt, step by step, to the most remote corner of the Kingdom. Migrants enumerated in a certain centre of absorption will consequently grow less with the distance proportionately to the native population that furnishes them.
3. The process of dispersion is the inverse of that of absorption and exhibits similar features.
4. Each main stream of migration produces a compensating counter-current.

5. Migrants proceeding long distances generally go by preference to one of the great centres of commerce and industry.
6. The natives of towns are less migratory than those of the rural of the country.
7. Females are more migratory than males over short distances.
8. The incidence of migration increases with increasing technological development.

Subsequent studies have shown Ravenstein's broad generalisations to be basically correct and many of his concepts have been refined and expressed in more sophisticated forms. For example, the Inverse Distance Law, expounded by G.K. Zipf, states that 'the volume of migration is inversely proportional to the distance travelled by the migrants.' This may be expressed mathematically as

$$N_{ij} \propto \frac{1}{D_{ij}}$$

Where N_{ij} is the number of migrants from town i to j and D_{ij} the distance between the two towns.

A number of studies have not only examined the relationships between volume of migration and distance, but have also considered the influence of 'opportunities'. In 1940, S. Stouffer proposed his Theory of Intervening Opportunities, in which he suggested that 'the number of persons going a given distance is directly proportional to the number of opportunities at that distance and inversely proportional to the number of intervening opportunities'.

$$N_{ij} \propto \frac{O_j}{O_{ij}}$$

Where N_{ij} is the number of migrants moving from town i to j, O_{ij} the number of opportunities at j, and O_{ij} the number of opportunities between i and j. Stouffer defined 'opportunity' in terms of vacant houses, but others have interpreted it in terms of employment opportunities. Stouffer later refined his model by the addition of another variable, that of 'competing migrants at destination'.

Another interesting development has been the application of Gravity models to migration analysis. Ravenstein's first law of migration implies that distance is a barrier to movement and that long-distance moves require exceptionally strong attractive forces. It may be argued, therefore, that variations in the size of destination centres will influence the strength of their attraction. At the same time, the volume of migration decreases with increasing distance from source, so that a nearby city may be expected to have a stronger attraction than a distant one of equal size. Such a situation may be expressed as

$$N_{ij} \propto \frac{P_i \times P_j}{D^a_{ij}}$$

where N_{ij} is the number of migrants moving from town i to j, P_i the population of town i, P_j the population of town j, D_{ij} the distance between towns i and j, and a is an exponent of distance.

According to the formula, the volume of migration between two towns j and j is directly proportional to the product of the two populations and inversely proportional to an exponent of the distance between them. Thus, if we consider migration to London from Reading, Leeds and Cambridge it might be expected that the volume of the three migration flows would be in the ratio of 2-1: 2-4: 1-2.

	Reading	Leeds	Cambridge
Population (in thousands)	131	744	106
Distance from London (km)	61	306	87
	2.1	2.4	1.2

If the actual value of any one of the migration flows were known it would be possible to estimate the other two. Numerous assumptions are, of course, implicit in such a model; for example, that apart from the friction of distance there is equally unimpeded movement between each source and destination, and the idea that members of each of the sending populations are equally attracted to the destination. In reality, as Stouffer has shown, variations in the availability of housing and employment complicate the model. Nevertheless, the basic formula can be weighted in various ways to give a closer approximation to reality. An example is given below.

$$N_{ij} \propto \frac{XP_i \times YP_j}{D_{ij}}$$

Where X and Y might be values of 0-5 and 1-0 respectively to take into account the different levels of attraction of the two towns, low at i and high at j.

It could be argued that inequalities of one type or another are virtually preconditions of migration. Such an idea underlies the 'Push-Pull Theory', which has long been implicit in much of the writing on migration. 'Push' factors are those influences which are thought to initiate migration flows; they are the adverse conditions which cause population to seek a living elsewhere and include, among others, low wages, poverty, unemployment and natural disasters. 'Pull' factors are the attractions, real or imagined, of destination areas: they include high wages, cheap land, attractive living conditions and opportunities for economic advancement.

Many recent studies of migration have tended to accept a general background of 'push-pull' factors, but have also pointed out that any migration flow may also be interpreted as the aggregate result of numerous personal decisions about whether or not to move, when to move, and choice of destination. Thus, emphasis is increasingly being placed on the psychology of decision-making. D. J. Bogue, for example, has discussed migration-stimulating situations and examined the factors influencing choice of destination. One problem in such an approach is the fact that the decision to migrate may not necessarily be a rational one.

At the present time many studies are concerned with the construction of mental maps which attempt to represent the perceived environment rather than the real world situation. As pointed out in Chapter Two, the mental picture of an area may be of more geographical significance than reality. Such a line of argument is particularly relevant in migration studies, where choice of destination may be strongly influenced by the migrant's impression or perception of an area.

Types of Internal Migration

Detailed descriptions of the various types of internal migration lie beyond the scope of this book, but it may be of interest to draw attention to certain typical and recurrent types of internal movement. The importance of these various categories of movement depends to a large extent upon the stage reached in the social and economic evolution of the country concerned. For example, the development of rural settlement often shows phases of advance, stability, retreat and reoccupation. Similarly, rural-urban migrations often involve a phase of rapid urbanisation or concentration of population followed by a gradual diffusion from the urban centres.

Reference was made in chapter five to the ecumene—nonecumene concept, and it was suggested that an advancing settlement frontier was characteristic of earlier times. At present, extension of the inhabited world proceeds only slowly. Examples of recent pioneer advance may be noted in parts of the Soviet and Canadian Arctic, in certain desert areas such as the Negev, and in parts of the equatorial forest such as the Amazon Basin. Successful new settlement in such areas is inevitably government sponsored, for at the present time the individual pioneer is rarely able to confront nature singlehanded with success.

More typical at the present time are the numerous examples of rural depopulation. Decline of rural settlement by migration loss is generally most pronounced in the isolated or peripheral districts of a country or in areas of exceptional physical hardship. Under these circumstances the 'push' stimuli operate particularly strongly. The process may be well illustrated by reference to the North-West Highlands of Scotland, the high mountain valleys of the Alps, or the skerries and islands off the Norwegian coast; all are characterised by declining populations, largely as a result of a long-continuing negative migration balance. Because of the age-selectivity of migration, the process is cumulative. The younger sections of the population tend to move away, with the result that birth rates fall and death rates rise, thus accentuating the simple loss of numbers by migration.

The bulk of these rural migrants eventually find their way to the towns. Indeed, the process of urbanisation is the most powerful and universally dominant form of internal migration at the present time. New towns are being created, large cities are becoming ever larger and groups of towns and cities are merging to form conurbations. The processes involved will be examined more closely in Chapter Seventeen. At this point it is sufficient to note certain of the results of the urbanisation process. In the UK, for example, the 1981 census classified 77 percent of the population as urban. A comparable proportion of urban population is found in many other countries, including Australia, Israel, the Netherlands and Belgium. In Japan the proportion of town-dwellers rose from 22 percent in 1925 to 76 percent in 1980; the USSR from 18 percent in 1926 to 65 percent in 1984; and in France from 46 percent in 1921 to 73 percent in 1982.

What, if any, are the limits to this process of urbanisation? Statistics indicate that in the case of many cities the limit may already have been reached. There is evidence to suggest that the congestion, overcrowding, inflated land and property values and health hazards of the inner districts of many cities have generated a movement of decentralisation or suburbanisation. A large proportion of those who actually work in the inner city can now only afford to purchase a house in the outer suburbs. Thus, between 1971 and 1981 all the Inner

London boroughs maintained an established trend and showed a loss of population, while during the same period the outer suburban districts were all distinguished by vigorous population increase. A similar decentralisation is evident in most major cities in the UK and is indicative of the latest phase in the changing pattern of internal migration.

Many characteristic features of internal migration are revealed, which shows the inter-regional balance of recent migration in Norway. North Norway, the region of greatest hardship, is characterised by a negative migration balance and loses population to all other regions. Conversely, the main centre of attraction is the country's most highly urbanised region, centered on the capital Oslo. A recent census revealed that over 60 percent of Oslo's population were not born in the city, but had migrated there from other parts on Norway. Much of this drift to south-east Norway is in stages via other urban centres such as Bergen and Trondheim.

International Migration

International migration is becoming a more visible and important issue in international relations and in national self-concepts.

Globally, the number of international migrants increased from 75 million to 120 million between 1965 and 1990, keeping pace with population growth. As a result, the proportion of migrants worldwide has remained around 2 percent of the total population. In 1990, international migrants were 4.5 percent of the population in developed countries and 1.6 percent in developing countries.

These global estimates mask important difficulties in measuring migration. Only a handful of countries regularly count inflows of foreigners and returning citizens, so it is virtually impossible to make estimates of foreign-born migrants except via periodic censuses. Migrants sometimes avoid or are neglected by census-takers, and they are countred or classified in different ways by different countries. Migration is often the result of conflict, persecution or weather-related hardship, and as a result it fluctuates greatly from year to year and may be accompanied by chaos, making precise counts difficult.

Virtually all countries have been the destination of some migration in this century of rapid and universal transportation. Recipient countries for migrants have become more diverse since 1965, both in terms of the number of migrants they receive and their share of total population. The number of countries with a migrant population of 300,000 or greater increased by more than 50 percent between 1965 and 1990.

The percentage of women migrants has increased in recent decades, to 48 percent of all international migrants in 1990. Most women who migrate for employment tend to be concentrated in low-status jobs, and many are particularly vulnerable to expoilation and harassment. The globalization of capital and trade flows is causing unpredictable changes in the fortunes of developing countries, as investment capital rapidly moves in and out of fragile economies. In turn, these movements drive both internal and international migration. The growing informalization of the economies of many countries has also intensified the interaction between irregular employmentand irregular migration.

Increased immigration has been recommended by a number of demographers and economists as a means of balancing the effects of fertility decline and the resultant ageing of the population. For instance, a labour shortage in Japan has been met by expanding the number of foreigners (including descendants of former Japanese emigrants) who can be admitted to the country. Between 1985 and 1995 , the legally resident foreign population in Japan increased by 60 percent, and the number of undocumented aliens also grew. The economic effects of migration run both ways. Throughout the world, remittances by migrants from more-to less-developed countries remain an important mechanism through which international migration influences development.

Population Growth

There are three ways in which a population can grow: when natural increase exceeds net emigration; when net immigration exceeds natural decrease; and when there is both natural increase and net immigration. But the factors which affect the contributory variables of mortality, fertility and migration are extremely complex and are as yet little understood either individually or in combination. Why do some populations increase faster than others? Why do particular groups of females have more offspring than others? How are diseases communicated and what is their influence in restraining the growth of population? These and other similar conundrums are of fundamental importance for the study of population structures. Attempts to solve them require the construction of models and theories which are intended to help in the process of simplification and explanation. Because of the importance of theory construction for the analysis of population structure and distribution, it will be necessary to discuss some of the different ways of developing theories, their strengths, weaknesses and complexities, before the main question to be posed in this chapter can be tackled, namely, 'How and why does population grow or decline?'. The results of this general discussion may then be applied to the specific problem of constructing theories to explain population growth.

Systems and Theories

It must not be imagined that the characteristics of a population are the simple direct outcome of processes which can be accounted for in terms of a variety of causes. The quantitative and qualitative aspects of a population will have reciprocal influences not only on other populations, but also on individual members of the same population. For example, economic development in human societies can be both enhanced and dampened down by rapid population growth, whilst in animal populations chronic overcrowding can initiate genetic mutations and even lead to reduced fecundity. The interplay of factors which cause population growth to occur and the influence of that growth on the organisation of society make a full understanding of the demographic system unlikely. As a consequence of its complexity it is necessary to 'break in' to the overall system at some specific point and to deal with sub-systems which comprise only a limited number of interacting variables. The two main sub-systems will therefore focus on 'what causes population to...' and 'what influence population has on...' population as effect and cause. However, within these sub-systems there are yet more associations which need to be isolated. Change in fertility could be taken as the definitional criterion for a sub-system, for instance, which would make the pattern of fertility the outcome to be explained by its associations with a number of other causal variables within the sub-system. In this case, the effects of changes in fertility on the organisation of society or on the other demographic factors would be temporarily removed from discussion. Apart from this 'systems closure' approach it is also common for variables to be 'held constant'. For instance, the influence of migration can be removed while the interaction of mortality and fertility is studied.

Once a system of enquiry has been identified, it becomes important to establish the level of generality which is to be aimed for and to select a methodology by which that aim may be realised. What kind of theories are required? How are they to be constructed? These are the questions which need to be resolved.

Most social scientific theories have the following form: they contain *theoretical statements* from which *empirical statements* can be logically deduced and which can be related to observations; that is[they are verifiable. Causal theories contain theoretical statements about the association between two or more *variables,* variables which have *observable values.* For example, the three variables *A, B* and *C* may be connected in a variety of ways. If A causes B and B causes C then the following representation is appropriate.

$$A \rightarrow B \rightarrow C.$$

Similarly, A and B in combination may cause C

$$(A + B) \rightarrow C$$

or A and B could be causally related to C, but unrelated to one another.

$$A \rightarrow C.\ B$$

Clearly these are very simple examples yet they all rely on the term *cause* (→) for their meaning. *Cause* can be used to refer to the necessary and only pre-condition for the existence of a phenomenon; a sufficient condition; or even a contributory condition. In social scientific enquiry it is unlikely that any meaning other than the last-mentioned will be an unrealistic one, for associations and connections which are in reality like those shown above are trivial. Causal theories have to be constructed using proxy or surrogate variables which are contributory, although necessary, pre-conditions for the existence of a phenomenon. A and B must stand in for D to Z, some of which may not as yet be identified as having connections, or if identified they may not have observable values. Despite problems encountered in the definition of *cause* social scientists have extended the range of empirical statements to include positive, negative and two-way associations between variables. For instance, although A is a cause of C the observed value of C is inversely related to that of A, whilst B and C are causally and reciprocally interrelated in a positive fashion, thus,

Such causal empirical statements have been elaborated to include whole series of causal variables whose relationship with other variables can be hypothesised. The following gives an example, which can also be written in matrix form,

To

Causal link		**A**	**B**	**C**	**D**	**E**
	A		—			—
	B				+	+
From	C		+			+
	D			+		
	E					

Here all the variables, but *A,* are in some way causally related to other variables, although variable *E* is the destination of the most causal links. Much more sophisticated causal empirical statements have been developed by social scientists, but all are in some sense the product of systems closure and the holding of certain variables constant. The above brief outline raises a number of additional questions, three of which will be posed here. How may empirical statements be verified? Are there different kinds of theoretical statements? Where do the theoretical statements come from in the first place? The sociologist Hubert M. Blalock (1961, 1969) has offered a means of translating empirical statements about variables whose values are *measurable* into statistically testable statements. For example, the above matrix expresses the relationship.

$$E = f\ (A.B.D)$$

which it is possible to write in terms of a multiple regression model

$$E = \alpha — \beta_1 A + \beta_2 B + \beta_3 D + e.$$

Here one takes *E* to be the dependent variable; it depends upon three independent ones *(A,B,D)* but is inversely related to *A* and positively related to *B* and *D*. The step from the empirical statement, which is embodied in the matrix of causal links, to the additive multiple regression equation is not a straightforward one because it requires that a number of assumptions be made. Causality and dependence are not synonymous terms, indeed the multiple regression equation does not show causation, only statistical association. Further, the general linear model, upon which multiple regression analysis is founded, requires that variables should have measurable values which must in turn obey certain statistical rules (see Johnston, 1978). This said, multivariate analysis does provide a means of establishing whether dependency links exist between variables, the signs of those links, and even their strengths; and since it is illogical to expect causation without dependence at least part of the verification process is achieved by its use.

Although variables may have observable values these need not necessarily the measurable. In consequence statistical testing is often an inappropriate means of verifying empirical statements. Many human characteristics which are legitimately the subject of theoretical statements are essentially qualitative. People's attitudes, superstitions, fears and desires are crucial aspects of any interpretation of their behaviour yet these are also rather nebulous attributes which defy both precise definition and measurement. Similarly, many of the 'world theories' of the social sciences contain variables which are themselves whole theoretical systems. The term *modernisation* epitomises this problem; it implies material, ideological, economic, social and political changes which are themselves the focus of complex theoretical statements. Many 'grand theories' of social change are expressions of belief rather than the source of testable empirical statements, yet it is certain that the quality of derivative theories would be much reduced without, for example, the imagination of Freud or the historical vision of Marx.

One way out of this maze would be to follow Robert K. Merton and attempt to develop *theories of the middle rage*. These are not all-embracing unified theories which explain all observed regularities nor are they propositions of the most unsophisticated sort, rather they involve abstractions which are, 'close enough to observed data to be incorporated in propositions that permit empirical testing' (Merton, 1967:39). These middle range theories, Merton advises, should be *special theories* applicable to limited conceptual ranges which can gradually be merged to create even more generalised grand theories. Once we readopt the empirical verifiability criterion for the construction of theories we are able to base the verification process on the foundation of statistical testing, provided we are willing to deal with dependence rather than cause and can measure as well as observe.

The problem of theory verification has, however, been cast in a rather different mould by the work of Karl Popper. He emphasises the role of falsification—theories are to be rejected if they can be falsified although only temporarily accepted if they cannot (see Poper, 1965; Harvey, 1969:39; Geogory 1978:35). Theories must therefore be tested although it is not always clear whether as a result one has falsified or temporarily accepted. *A* causes *B*—the introduction of a family planning programme causes fertility to decline—is a theoretical statement which can be tested empirically, it is falsifiable, and indeed has been found to be false in the case of the Nigerian population (Caldwell and Caldwell, 1977; Caldwell and Ware, 1977), but in most instances the statement is valid, if not very helpful. To reject on the strength of one counter-example would be in this case to fly in the face of common sense, although even this objection seems to be controversial (Popper, 1972: 32-105). Popper's falsifiability criterion does pose new and practical problems for the social scientist, but its emphasis on how the mechanism of theory testing should work is nonetheless important.

Hay (1979) has taken a different tack by suggesting a means of assessing whether a theory is *useful* or not. Hay's criterion is based on a positive answer to the question, 'Does this theory contribute an explanation of a part of the observed variation which would otherwise remain obscure?' in contrast to a question like, 'Does this theory *totally* explain the observed variation?' (Hay, 1979:9).

The verification issue also has disciplinary and ideological elements. Contrast the views of development economist Leibenstein and sociologists Hindess and Hirst. Leibenstein (1954:5) argues that, 'one of the primary objectives of theory construction is the careful formulation of a set of interrelated ideas from which it is possible to deduce meaningful theorems, by which we mean simply a body of conceivably refutable propositions'. This should lead to, 'one or more non-trivial theorems or propositions that can be put in such form as to be conceivably falsifiable by empirical research'. But to Hindess and Hirst (1975:3), 'Our constructions and arguments are theoretical and they can only be evaluated in theoretical terms—in terms, that is to say, of their rigour and theoretical coherence. They cannot be refuted by any empiricist recourse to the supposed "facts" of history' (but see Hindess and Hirst, 1977:41).

The question 'How may empirical statements be verified?' is therefore an extremely complex one. Its answer depends upon the form of the theory under discussion (nature of the variables etc.), and the rigour of the criteria one wishes to use, for instance, in translating cause into dependence. Answers to this question will vary depending on the kind of theory which is being constructed. Whilst the physical sciences tend to deal with deterministic theories the social sciences are usually relegated to the imprecision of probabilism. Empirical statements are really probability statements which rely for their verification on the observation of the joint occurrence of measurable veriables. This is the inductive-statistical approach to theory building; it is the cornerstone of positivism in geography and as such it has become the target of many hammer-wielding critics. Two points regularly made by these critics questions, firstly, the logic of inductive reasoning that moves from the particular to the general usually by repeated observation or experiment (*i.e.*, although in n trials $A \rightarrow B$ it need not necessarily do so in the n + 1th trial) and, secondly, that these inductive-statistical theories are counter-revolutionary. This point is made by Harvey (1973: 120-52) and illustrated by reference to the Negro ghetto. Most geographical theories relating to ghetto formation are of the inductive-statistical form. They attempt to identify the causal links which make ghettoes as they are, the association between the causal variables being implicitly assumed to be invariant and in consequence the ghetto being the only possible outcome. In contrast, a revolutionary theory seeks to establish a network of variable association which would lead to the destruction of the Negro ghetto. Emphasis is transferred from theorising over what *is* to that which *should be;* the ethical neutrality of interpretation is abandoned for the commitment to change.

The kinds of theories mentioned above are not exhaustive, but they do indicate the variety that is available to the theorist. His choice of approach will be conditioned by the particular problem at hand, by the tradition of his intellectual discipline and even by his own personality. However, the question 'Where do the theoretical statements come from in the first place? still remains. As usual there is no one answer. Theoretical statements are rarely divinely inspired today. Most are born of intuition, speculation, or from the reconstituted debris of theories which have been rejected or only temporarily accepted, or have been thought insufficiently revolutionary. Theory construction, although somewhat diffuse, can follow certain general guidelines which counsel the listing of relevant variables drawn from any one of the above sources; their precise definition in an observable or, ideally, a measurable form; the elaboration of an empirical statement which will contain casual links between the variables; and, finally, the specification of a method of statistical testing, falsification or critical assessment which will enable one to judge the value of the particular theoretical statement as an intellectual tool.

The theories constructed and employed by population geographers have tended to be of the inductive-statistical variety and as such they have usually specified the dependency relationship between measurable variables. 'Migration between two places depends upon the characteristics of the interacting places, the intervening places and the potential migrants themselves' is one example in which once the characteristics have been defined precisely the resulting empirical statement can be tested using multivariate analysis. Such statements are typical of the use of middle-range theory in geographical studies. However, population geography has also taken to itself at least one grand world theory, namely that of the demographic transition which deals with changes in mortality, fertility and migration both over time and through space. It is a theory which is only partially testable because it not only contains statements about the past and the present, but it also predicts changes for the future. Verification also raises problems for behavioural theories in population geography. This category of theories deals with the association between attitudes, behaviour, and the organisation of society which is often observable in spatial patterns (*i.e.* at its simplest the $A \rightarrow B \rightarrow C$ causal chain). It has proved particularly valuable in studies of the motivation for migration and the links between prejudice, discrimination and spatial segregation, but because it deals with both psychological and sociological variables it tends to express the observable rather than the measurable. Nevertheless, these theories are crucial for the understanding of the behaviour patterns of individuals in population geography, even though they have to be critically assessed rather than statistically tested (White and Woods, 1980). During the 1970s the construction of revolutionary theories was begun in population geography. Most of these attempted to deal with the vexed problems of distribution of and access to resources, overpopulation and underemployment. Harvey (1974), for instance, has discussed the population-resources problem and the contribution to the debate made by Robert Malthus, David Ricardo and Karl Marx. Only Marx's dialectical materialism, Harvey argues, provides a dynamic analysis which is explicitly ideological. Here the emphasis is on altering the 'social organisation of scarcity' rather than reducing numbers. Although Harvey's (1974) analysis is destructive—it is unclear how the implications for distribution are to be effected—his work has placed both the supposition of ethical neutrality and the new-Malthusianism in sharp perspective and has cleared the way for a constructive revolutionary theory of population-resources.

It is evident from the above that population geographers in their attempt to construct theories have taken to heart Wrigley's advice that, 'Perhaps the most sensible attitude now as at other times to adopt towards the question of method in geography is to be eclectic—to use whichever method of analysis... appears to offer the best hope of dealing with the problem in hand' (Wrigley, 1965:17).

The central question to be posed in this chapter is, 'How and why does population grow or decline?' The task of providing an answer will be approached with the aid of the guidelines described above. Controlling factors will be discussed, their definitions and association examined as the first stage in the construction of theories of how the size of populations change over time. However, growth or decline is not simply monotonic; there have been cycles of population change which will require separate theoretical treatment. Finally, we shall return to the work of those two most influential of nineteenth-century social scientists—Malthus and Marx—to see whether they were able to provide satisfactory answers to the question.

Factors Controlling Population Growth

Whether a population grows or declines is controlled by the relative balance of mortality, fertility and migration, which are in turn influenced by six groups of factors: biological, environmental, economic, social, political and technological (see Sauvy, 1969).

Biological factors are responsible for both a population's physiological and epidemiological characteristics. The maximum number of offspring a female can have is determined by biological factors, as is the sex ratio at birth. The age at menarche and menopause controls the number of years a female is potentially fecund, whilst the proportion of potentially fecund who are actually sterile is also largely the result of biological influences. Physiological variables establish maxima and minima between which the normal pattern of reproduction is to be found (see Parkes, 1976; Cavalli-Sforza and Bodmer, 1971).

The aetiology of diseases is an important influence on mortality patterns. Man's ability to first understand and then control the incidence and impact of infectious diseases have been relatively new found skills. For example, bubonic plague ravaged the population of western Europe for over 300 years without being scientifically explained whilst in the nineteenth century the significance of tuberculosis declined seemingly without the aid of medical practice (McKeown, 1966). The dramatic impact of newly imported diseases on a population lacking any resistance has often been recorded, but it was particularly important as an adjunct to the colonisation of the New World (McNeill, 1977). Certain epidemic diseases have therefore had a significant influence in creating demographic crises, whilst the endemic nature of others in particular areas, has led to the maintenance of relatively high mortality. Smallpox and influenza are typical of the former class of disease, cholera and malaria are examples of the latter.

Whilst the biological factor is particularly important for its influence on reproductive performance and mortality conditions, the environmental factor also influences the propensity to migrate. An examination of a map showing the distribution of world population makes it clear that there are some very negative areas for human settlement—mainly those areas experiencing very high or very low temperatures in the tropical and polar regions—but that most of the rest of the globe's land surface is well populated. Obviously there are differences between Kansas and Bengal, but there are economic, social and historical reasons for them and not simply environmental ones. Climatic fluctuations and the vicissitudes of whether have also has substantial indirect influences on population growth, particularly by generating short–and long-term crises in populated areas (Lamb, 1972, 1977: 423-549). Subsistence crises in particular have been related to harvest failures which are themselves a function of late springs, wet autumns or seasonal droughts (Le Roy Ladurie, 1972). During such crises mortality, especially amongst the young and the old, is usually heightened, rural emigration is accentuated and fecundity is often reduced.

The environmental factor is also responsible for a number of other hazards which can have locally devastating consequences. Of these flood and earthquake are possibly of most importance. The Hwang Ho, for instance, has gained a notorious reputation as 'China's sorrow' and despite centuries of canalisation 10 million people lost their homes and 50 milion were directly affected by the floods of 1931 (Kolb, 1971: 109). The environment imposes more subtle checks and barriers such as the limits on the growth of crop varieties which indirectly influence the level of rural population density attainable; rice and wheat cultivation provide a contrast on this point (Braudel, 1973: 66-120). The natural environment also restricts and channels the movement of people, information and the spread of technological innovations (Hagerstrand, 1967). In short, although one would not wish to return to using the concepts of the environmental determinists it is undoubtedly true that the character and vagaries of the natural environment do impose constraints, inflict unpredictable dangers and provide opportunities to which man has adapted himself, succumbed and taken advantage to varying degrees at different times and in different places.

That human societies are divided into 'classes', 'status groups' and 'parties' was one of the most important contentions of the German sociologist Max Weber (Gerth and Mills, 1948: 180-95). Although this basis for stratification has not been accepted wholeheartedly (see Runciman, 1968; Freund, 1968: 167) it is important in focusing attention on the economic, social and political dimensions of social organisation. Whilst no one, including Weber, would claim that these strata are mutually exclusive one is forced to recognise that the economic relationships which create classes—the disposal of goods or skills for the sake of income—are not the same as the ascriptive processes of prestige allocation which create distinctive status groups, or the mechanisms which determine the distribution of political power in society. Economic, social and political factors are essential, although interrelated, aspects of the wider demographic system and as such they must be treated as distinct and separate influences.

The time-honoured adage that population grows is response to the demand for labour has all the characteristics of an overworked half-truth: it is simple, eminently reasonable, yet presents only one side of the problem. Certainly, the massive demand for labour by the expanding American economy in the nineteenth century did lead to emigration from the Old World just as the industrial revolution in Britain created the opportunity for rural—urban migration, but there can be population growth without there being a demand for labour, whilst labour shortages can be chronic without inducing population growth or redistribution (Simon, 1977). The essence of the argument is that one is to regard fellow human beings, and even one's as yet unborn offspring, as though they are commodities—consumer durables, for instance—the supply of which will be adjusted to demand, even if this mean the creation of population surpluses or shortages and the necessity for transport. Who one is will be very important here.

The economic factor also has a considerable, if indirect, influence on population growth through the standard of living. The inverse relationship between improving living standards and mortality has often been observed, particularly as it works through better diet, clothing and housing to ameliorate the worst excesses of those diseases which stem from undernourishment and poverty. Fundamental changes in the economic system associated will the agricultural, manufacturing and tertiary stages of development put strains on the structure of population as it adjusts from rural to urban and from manual to cerebral styles of life and modes of employment. These pressures may have influences on the desire to regulate fertility and certainly are important for generating both higher population turnover and longer distance 'betterment' migration.

It is not always easy to make a precise division between the influence of economic and social factors. For example, although 'class' is usually defined in practice in terms of occupation, the resulting combinations are placed in a hierarchy defined by levels of prestige which makes them 'status groups'. The terms 'socio-economic groups' or 'social classes' are therefore used to express the ideology that sub-populations with similar occupations and degress of prestige have similar life styles and consumption patterns; have the same attitudes and behave in a similar fashion; and that inter-class is significantly greater than intra-class variation. The social class factor is most important in accounting for differential population growth. Mortality, fertility and migration patterns are all empirically related to differences in class.

The pure social factor can be thought of as expressing a large variety of variables, the cultural, religious, linguistic, ethnic and racial characteristics of a population which can all be important in giving it a distinctive demographic structure. Such attributes are often used as signs of inferiority or superiority and are thus, potential means of self or imposed isolation, which can itself have a variety of demographic and spatial expressions. American Negroes and central European Jews, for example, are two populations with cultural and enthnic identities which have been used to separate them from white America and gentile Europe. The oppression involved has also established demographic distinctions between minority and majority. The relative standing of women, the form of marriage pattern; the organisation of family groups, whether nuclear or extended; together with the inheritance laws are all ways in which social roles are formed and norms established. These roles and norms can mean the difference between, on the one hand, a society having monogamous marriages that are contracted when bride and groom are in their late twenties or thirties, where nuclear families are formed which use primogeniture as the rule for the transfer of property and, on the other, a society with early marriage, extended families and partable inheritance. The fertilty of females in the first-mentioned society is likely to be lower than that of those in the second, although which of the contributing variables is the most important in creating the tendency remains to be seen.

The social factor is naturally an extremely important influence in both fertility and migration patterns. It embodies the system of values that groups use to order their day-to-day existence, together with the more intangible interpretations they may place on the reasons for that existence. Such group values place obligations and constraints on members which lead them to behave in a more or less similar way. For example, they force them to marry; have children, not too many but not too few; and to form separate households away from parents and in-laws. The individual may choose not to conform, but social ostracism might be the result of such defiance. In this context demographic behaviour can be regarded as but one part of the wider ambit of social behaviour.

The potential migrant is also susceptible to group pressures and restraints, since the mechanism of chain migration is an extremely powerful one which is even capable of generating a 'migrating mentality'. This force is not confined to truly nomadic populations, but is equally applicable to twentieth-century urban Americans and Caribbean islanders. In making migratory moves individuals often use their social networks to gather information about possible destinations. Once the decision to move has been made and the venue chosen they are also used to gain access to both accommodation and employment. The active migrants are then joined by passive migrants, who in time often establish an immigrant colony within which the values of the homeland are gradually adapted to the new social and economic environment with its consequent influence in changing demographic patterns. This sequence of events, whilst not universal, has been observed frequently enough to provide a valuable example of the close interrelationship between social, economic and demographic variables, all of which change during the sequence.

The political factor is also an expression of an assortment of variables which are connected with resulting population patterns and with other causal factors. Nationalism, for instance, cuts across geographical, cultural and class barriers; it creates sentiments of an all-pervading nature and establishes the most significant of economic units. The distinctive government policies enacted in these separate nation states are bound to influence the size and structure of the populations which inhabit them. Immigration can be controlled; settlement in the frontier encouraged; family planing promoted; a national health service established—all of which will alter the makeup and distribution of a country's population. Similarly, at the local scale planing authorities can radically influence the distribution of population by their control over the location of new employment, the housing market and the provision of education, medical and social services facilities. All of these components are regarded as a reflection of the distribution of power in society and its translation into a system of differential allocation of benefits, rewards and compensations.

The technological factor can be thought of as comprising four aspects—invention, practical development, adoption and mass use—which are, to varying degree, responses to stimuli from the other factors. Advances in medical, transport, agricultural and industrial technology have all contributed to rapid population growth in the past 200 years by, respectively, reducing mortality via chemotherapy; allowing large-scale international migration by passenger ship and train; by preventing subsistence crises; and by improving living standards. Medical advances in the field of oral contraception technology have meant that birth control can now be practised effectively by the many millions of women who wish to space out their children's births and to limit their completed family size. However, these developments have not been ubiquitous nor have they always been wholly beneficial. The ever-widening gap between the technologically advanced Western states and those in the Third World means that health services are differentially distributed and there are wide disparities in living standards, both having important implications for long-term, rapid population growth.

It will be apparent from the above that these six factors are by no means independent of one another. New contraceptive methods will not be widely used and thus, have an impact in reducing fertility, unless there are favourable social and economic conditions for their adoption, such that available methods are seen as the means of obtaining a desirable objective. Although long-distance migration is often spurred on by economic motives its mechanism is very much a behavioural one with social political and even psychological components. As a final example, it can be shown that mortality responds to what are often chance variations in the biological and environmental factors, but it can also be affected by man's attempts to directly lengthen his average lifespan and by the way in which he organises his socio-economic environment.

The Process of Population Growth

The second question that needs to be asked is, 'What mechanisms are responsible for population growth?' The inductive approach is at its most vulnerable here because of the equifinality problem. Given that a logistic curve fits the growth of a particular population over time, one is left to infer what processes could have caused that pattern to be established, and since a number of different processes can lead to similar outcomes it is difficult to be certain whether the causes isolated are in fact the most important. Where experimentation is possible, as in the case of *Drosophila melanogaster* mentioned above, one may be reasonably certain what mechanisms are at work because all the variables involved can be controlled and observation is relatively straightforward. Most animal populations pose far greater interpretative problems than *Drosophila,* whilst human populations represent the most complex of demographic systems.

The important mechanisms controlling *R* for animal populations have been outlined by Lack (1954, 1966) and Clark *et al.,* (1967: 33-

56). Following Lack (1966: 2-3) we may divide these into three. First, density-dependent factors are responsible for the maintenance of a degree of balance in numbers by leading to changes in the reproductive rate; the death rate, due to food shortages, predation or disease; and even to self-regulating behaviour associated with territoriality. Environmental change may establish new levels of *K* which will in turn influence these density-dependent factors to check *R* or allow it to increase. Such concepts are associated with the name of their most persistent advocate. A.J. Nicholson, but they have been summarised in a most succinct from by Lack himself. 'Birds and other animals can increase in numbers with great rapidity but are usually held drastically in check. Nearly all animal populations fluctuate irregularly between very restricted limits. Their comparative stability must be due to eontrolling factors which are density-dependent' (Lack, 1954: 20). Secondly, there are density-independent factors which are linked with short-term random changes in, for example, weather patterns. If these factors are of prime importance in regulating growth, as Andrewartha and Birch (1954) have argued, then the ability of members of a population to change the conditions in which they live by migration also becomes significant. Thirdly, density-dependent factors are taken to be the most important, but density itself is not regarded as beyond the control of certain animals which, by means of spatial dispersion and constraints on breeding, regulate their numbers below the limit set by food supply. The existence of social behaviour capable of such regulation has been propounded by Wynne-Edwards (1962, 1965). These three sets of factor combinations embody the range of influence which control animal population numbers. They all work through the balance of $(b-d+im-em)$, but lay relatively greater emphasis on *b, d* or *im* and *em;* and on external controls or induced adaptation. In Nicholson's theory *d* is of prime regulatory importance and is determined by food supply, although *b* can also very; for Andrewartha and Birch *d* is randomly influenced; whilst for Wynne-Edwards *b* can be lowered so that *d* will not oscillate violently.

Lack's (1966) own studies of bird populations have tended to show that the most important control on them is density-dependent mortality and that starvation outside the breeding season is particularly significant in this context. He comments that, 'when a bird is introduced to a new region and becomes established there it at first increases rapidly, which shows that it is capable of rapid increase, but eventually its numbers level off; which shows that the capacity for increase has been checked, evidently by factors which operate more strongly at higher than lower densities of population. Thereafter the introduced species, like those native to the region, tends to fluctuate, in most cases irregularly, between limits that are small compared with what is theoretically possible' (Lack, 1966: 3). Wynne-Edwards's ideas are treated with some sceptism and whilst random shocks can obviously have a great influence on numbers in the short-term it is pressure on food resources that is the dominant long-term regulator of population size. In general, therefore, density-dependent d is the most important control on the size of animal populations so that as p_t tends towards K, when $Pt<K$, d will be increased and R will be diminished. In practice it is also possible for $Pt > K$, in which case d will remain high and R will be negative until Pt ` K.

Although this description of animal population does not do full justice to the subtleties of the matter it nonetheless serves to illustrate how it is possible for such populations to approximate the pattern of logistic growth through the mechanism explicit in the causal chain: environment (food supply) $\rightarrow K \rightarrow R$. What are the processes operating in human populations? For example, why has the growth of the American population apparently approximated to the logistic form?

The birth, death and migration rates for the American population reveal that in this particular case net immigration has been a most important contributor to the rate of population growth, but that variations in both b and d have also had a substantial influence on R. (Bogue 1969: 127-46) provides a concise outline of the development of America's population, but see also the papers repriuted in Vinovskis (1979). During the late nineteenth and twentieth centuries the death rate fell persistently and significantly while the birth rate also fell to a low point in the 1930s, since which time it has oscillated about a relatively low level. Is this the result of density-dependent controls? In a sense one could argue that the immigration restrictions imposed in the 1920s were a response to perceived overcrowding, but the changes in mortality and fertility which are reflected in d and b certainly were not. Mortality fell because the private and public health of the population was improved, diet became better balanced and medical science was able to both cure and prevent many fatal illnesses (Hermalin, 1966; Rao, 1973; Preston, Keyfitz and Schoen, 1972). The variables influencing the pattern of changing fertility, are far more difficult to specify, but declining child and then infant mortality, rising economic and social aspirations, together with increased availability and effectiveness of birth control methods are factors which are normally thought to be influential (Whelpton, Campbell and Patterson, 1966; Ryder, 1969; Rindfuss and Sweet,

1977). The impetus to establish a low fertility pattern was more likely to have been the recent acquisition and desire to retain material benefits, than the pressure of poverty resulting from a clash between population numbers and resources. Once low fertility had been established then temporal and spatial patterns could emerge which were the product of short-term economic fluctuations.

That part of population growth in the United States which can be attributed to migration, owes much to the close inverse relationships which developed in the period before the First World War between trade cycles in the European and American economies. Westward international migration within the area of the Atlantic economy therefore became an important balancing mechanism for labour surplus and shortage (see Willcox, 1929, 1931; Jerome, 1926; Thomas, 1954). This simple relationship could of course be upset by major European disasters like the 'last great subsistence crisis' (Post, 1977) of the early 1800s or the 'Great Irish Famine' of the 1840s (Cousens, 1960; Kennedy, 1976), which heightened the significance of push factors regardless of the contemporary American condition. The growth of the American population since 1790 is ultimately a matter of restrictions imposed on massive immigration; voluntary restraints applied to marital fertility; and medical control being developed and adapted for the limitation of mortality. The combined effects of these restrictions restraints and controls began to be felt in the 1920s and 1930s when, *R* first started to decline. It is this reduction in *R* through lower *d, b* and *im* that gives America's population curve its logistic form. Although the processes involved in establishing this time series are very complex, it cannot be reasonably argued that they involve a true density-dependent, *K*-related element. The United States case is therefore a prime example of how the pattern of growth over time can appear to take a form which, if it had been related to another population, would have implied completely different casual processes.

The example of Japanese population increase, raises similar problems. The case seems to be a simple one of the exponential growth of a population that in the eighteenth and early nineteenth century was static. However, the processes lying behind this pattern are extremely difficult to unravel. The birth and death rates in Japan reveal rather different trends. The Japanese birth rate increased from the 1870s to the 1920s and then began to fall. The post-war 'baby boom' was only a temporary reversal and the most dramatic decline in *b* occurred during the 1950s. The death rate shows a three-stage pattern—increase, plateau, decrease—in which the most rapid change has occurred since the 1920s. In the Tokugawa era (1600–1868) national surveys of the commoner population were made, but estimates of birth and death rates have to rely solely on local village population registers. The family reconstitution studies which have used these registers have tended to show that in the eighteenth and the first half of the nineteenth centuries b and d both ranged from 0.015 to 0.030, while b occasionally exceeded 0.040 (see, for example, Hanley and Yamamura, 1977: 211; T.C. Smith, 1977: 40). The low levels of b and d suggest that the Tokugawa era in pre-industrial Japan was significantly different from the Meiji, since birth and death rates were both lower and tended to balance one another out. However, in the Meiji period birth and death rates were rising, but with b always higher than d. Since the 1930s and 1940s d and then b have both fallen back to, and in the case of d well below, average Tokugawan levels, even though overall values of R have not been altered by these modern developments.

Interpretations of the curves for b and d are many and various. One traditional theory stresses the impact of economic stagnation and famine in Tokugawan Japan which it links with the practices of abortion and infanticide. The rapid economic growth after 1868 is seen as the reason for the rise in the birth rate because it was associated with a growing demand for non-agricultural labour and an improving standard of living. In this interpretation population growth responds to economic change, but in a counter-theory presented by Hanley and Yamamura (1977) the control of population growth is seen as a necessary pre-condition for industrialisation, because when it combines with even slow economic growth it creates conditions in which living standards improve and *per capita* income rises. Although the validity of one of these theories has not been conclusively demonstrated there is evidence to suggest that villages in Tokugawan Japan were using infanticide as a means of controlling birth intervals, limiting family size and determining the sexual composition of families (T.C. Smith, 1977). This practice, together with abortion, meant that age-specific fertility was similar to, and in many-cases lower than, that experienced in contemporary European villages, where age at marriage was the prime indirect regulator of fertility. Social behaviour of this kind was certainly an important contributory factor in keeping the Japanese population static for over a hundred years.

Theories to account for the rising and falling curves of the trajectory are less controversial. Certainly the economic expansion of the Meiji period did permit controls to be relaxed and the birth rate to rise, whilst in the 1920s and 1930s government disapproval of birth control measures meant that the relatively high levels of *b* which had been reached would be temporarily maintained. Since the Second World War the birth rate has declined dramatically, due partly to the legalisation of abortion in 1948, but also to the increased availability of birth control methods and the growing desire for smaller families which accompanied general prosperity and female involvement in the labour force (Taeuber, 1960; Mosk, 1977, 1979).

The case of Japanese population growth over the last 200 years is therefore an interesting one, for the multiplicity of processes that contributed to the static and exponential stages are intractable and in consequence not fully understood. It illustrates well the point that population growth can remain exponential (*R* is virtually constant) when *b* and *d* are both increasing, both static, and both declining. Unlike the American case, international migration has not been of great importance nor has there been an initial period of rising birth and death rates. However, there are distinct similarities in the way *d* has changed since the 1930s.

The American and Japanese examples are only two from a large number that could be used to show the different forms of population change over time and to illustrate the plethora of mechanisms that create those changes. They demonstrate that any general answer to the question. "How and why do populations grow or decline?" is likely to be only a partial one; that location in time and space make the experiences of individual human populations unique; and that whilst it might be possible to say 'how' the combination (*b–d+im–em*) has given rise to *R,* theories that treat the 'why' element of the question are as yet simple in structure and relatively weak in explanatory power. Nevertheless, one is forced to face the additional problem that the representations of population change over time. The next section considers examples of these growth cycles and addresses itself to their general form and the processes that have led to their creation.

Population Growth Cycles

Following the work of J.D. Durand it is usual to represent the growth of world population by a smooth curve which in its final stage is exponential in form. However, it is more likely that world population has increased in a series of cycles and that at times there has been actual decline. World population growth since 400 BC according to Durand's (1967; see also Durand, 1977) estimates and compares them with those made by Biraben (1979). Similar patterns of cyclical growth have been illustrated by McEvedy and Jones (1978: 342-51) in their

Atlas of World Population History. They identify three cycles which they call *primary* (10,000 BC-AD 500), *medieval* (500-1400, and *medernisation* (post-1400). The years 500 and 1400 are used to symbolise periods of population decline when, McEvedy and Jones suggest, population growth had reached and even exceeded some upper limit which had in consequence led to the establishment of a dramatically higher level of mortality. The medieval prayer *A bello, fame, et peste libera nos Domine* reveals three of the ways in which these periods of crisis manifest themselves: war, famine and plague. The coincidence of events in Europe and China, together with the high proportion of the world total which the combined population of those two regions comprised, has tended to give the course of world population growth this distinctively cyclical form.

The growth of population in England and Wales also provides an example of cyclical change. Hatcher (1977: 71), for instance, estimates that the population in 1300 was similar to that in 1650, that is about 5.5 millions. Three cycles are also in evidence here. The first ends around 1400; the second continues until the eighteenth century; and we are still in the third today. Since we can assume that in general *im=em* and if it does not then *im<em* the form for the curve must largely be the result of the balance between birth and death rates (R=b–d). Apart from the latter half of the fourteenth century, the curve consists of alternating periods when *R* is approximately zero and when *R* is well in excess of zero-that is, it may be looked on as a series of connected logistic growth curves each of which represents a cycle.

In theory there are four main ways in which a logistic type cycle can occur. The birth rate remains static while the death rate declines and then increases *b* increases followed by *d* declines and is followed by *b* and *b* increases and then declines while *d* remains static.

The two post-1400 cycles for England and Wales are likely to have been caused by different *b* and *d* sequences. Whilst it seems clear that the last cycle—post-1700, it is difficult to infer what processes created the 1400 to 1700 cycle. In is thought that the fifteenth and seventeeth centuries experienced particularly high levels of mortality, whilst the sixteenth century has been recognised as a period of rapid population growth in England. During the sixteenth century the death rate may have fallen to a relatively lower level and life expectancy at birth may have been in the low forties, but there are also grounds for arguing that the birth rate rose to give a mean completed family size for women married before age 25 of over seven. The evidence for these assertions comes mainly from the results of family reconstitution studies which use Anglican parish registers for individual villages (see, for example, Wrigley, 1966a, 1972a). They rely on comparisons between conditions in the sixteenth and seventeenth centuries which of necessity ignore the first stage in the 1400-1700 cycle. In the seventeenth century mortality seems to have been substantially higher and fertility lower than in the preceding century. If this sequence of events is a realistic interpretation then a modified form would be appropriate. (These issues are discussed at length in Wrigley and Schofield, 1981).

The implications of this model go against the normal analyses of pre-industrial population which stress the controlling influence of short-term mortality fluctuations. Such interpretations would describe the growth phase of the cycle. However, any interpretation of pre-industrial population cycles is bound to be conjectural in the absence of comprehensive and reliable quantitative evidence. This said, it is clear that for the growth cycles to have occurred there must have been oscillation in *b* or *d,* or both. There must have been stretches of good years when plague and other epidemic diseases were not as rife and periods of good harvests when agricultural improvements were made and when new land was brought into cultivation, when the production of food was adequate or even abundant. It would appear that the thirteenth and sixteenth centuries in England and Wales were such times. These 'times of feast' were also suitable for earlier marriages which in turn meant that the birth rate increased. The delaying of first marriages for females to the mid- or late-twenties was one of the most effective means of social control that pre-industrial societies could exert on their fertility and as is becoming increasingly more obvious, one which was used to good effect in western Europe from at least the sixteenth century onwards (see R.M. Smith, 1978; and more generally Dupaquier, 1979).

The 1400-1700 population growth cycle in England and Wales is, therefore, a difficult one to explain. The mechanisms which caused its first and third stages were essentially dominated by high death rates with lower life expectancy. Although the second stage was partly a response to falling *d,* it was also, and perhaps most significantly, influenced by rising *b.* The form and underlying mechanism of the post-1700 cycle are fundamentally different. In this cycle new levels of fertility and mortality have been reached in the twentieth century which are inconceivable in any former period. However, even the mechanisms underlying these changes are highly controversial.

The course of change in birth and death rates since vital registration began in 1837 is known. But it has been suggested that the period of rapid population growth which began in England and Wales in about 1750 was induced by a rise in the birth rate. The major protagonists for this point of view have been Krause (1967, 1969) and Habakkuk (1972). Their argument is that the process of industrialisation and economic expansion led to a demand for labour which in turn created conditions favourable to a reduction in the age at first mariage and hence a rise in the birth rate. Krause (1969: 127) refers to the view that the death rate declined in late eighteenth-century England as a 'statistical mirage', but there are many who, following G.T. Griffith (1926), have taken the view that the current cycle of population growth in England and Wales was initiated mainly by a fall in the death rate. Few would now agree wholeheartedly with Griffith that this secular decline was linked with medical advances and the establishment of hospitals. Razzell (1965) argues that inoculation against smallpox led to a significant mortality decline, whilst McKeown and Brown (1955) stress the importance of rising living standards and improvements in diet (see also Razzell, 1974). Whilst it seems that a major cause of the take-off of population growth in this period was declining *d*, a view endorsed by Lee's (1974, 1978) demographic simulations, it is also more than likely that *b* rose at the same time, a point which has emerged from reverse projections undertaken by members of the Cambridge Group for the History of Population and Social Structure (Wrigley and Schofield, 1981).

If the above account of the pre-vital registration phase of the current population growth cycle is based on estimates which are made like 'bricks without straw', as Glass (1965a) has remarked, then the construction of theories to explain the post-1837 aspect of the cycle ought to be more straightforward. It is not. We may follow Mckeown (1976, 1978) and argue that mortality fell in the second-half of the nineteenth century largely because of improvements in living standards and particularly diet which affected the incidence of such infectious diseases as tuberculosis, but it must be remembered that only 30 percent of the decline in mortality from 1848-54 to 1971 occurred before 1901 (Woods, 1979: 69), and that for certain groups, particularly infants, mortality did not even begin its secular decline until the twentieth century. Our analysis of mortality decline in England and Wales, together with that already mentioned, must take into account the important roles of chemotherapy, advances in surgery and anaesthetics, and the development of efficient medical institutions for the cure and prevention of illness. These advances have had a major impact on the population of Western scientifically advanced countries since the first advances were made in germ theory in the 1880s. By the 1980s in England and Wales life expectancy is approaching its biological maximum, so that mortality is likely to remain extremely low in the future.

Theories to account for the decline of fertility in England and Wales must be highly sophisticated to be at all realistic. Most of those put forward have stressed one or more of the following: the significance of relative economic pressure on the middle classes; the increasing availability of mechanical and subsequently chemical means of birth control; the declining economic value of children; and the growing economic value of women's time as more and more of them enter the labour force.

But despite rigorous analyses of class, regional and socio-economic fertility differentials there is as yet no single comprehensive theory to account for the changes in b which have brought about the final stage of the post-1700 cycle (see, for example, Innes, 1938; Glass, 1938).

The examples used here to illustrate the phenomenon of growth cycles suggest that there is no one guiding principle which can be called upon to account for the onset and subsequent cessation of rapid population increase. Clearly the density-dependent mechanism, which affects the regulation of animal numbers and generates the often observed logistic curve, does not have the same direct relevance to human populations which have a far greater degree of control over their own food supply, environment, fertility, mortality and mobility.

Without this control animal populations are unable themselves to alter K which means that natural, or man-induced, environmental change must be the ultimate arbiter of the level to which population can rise. Although Western man now possesses this controlling ability his power has by no means always been so profound. For this reason it is quite conceivable that there have been periods of 'ecological climax' when available resources were unable to provide for the needs of populations, but it is also the case that there have been periods of heightened disease activity which were unrelated to population size or environmental conditions. Under these circumstances the pleasing regularity of line which population growth cycles display is entirely deceptive. Their causes are most complex depending as they do upon particular combinations of the factors.

The vague and extremely imprecise models and theories which have been invoked above to help answer the question 'How and why does population grow or decline?' must be specified far more precisely before they can be tested, evaluated or rejected.

The inductive approach and emphasis on the reverse link between form and process do provide a rudimentary starting point. Subsequent chapters will make the additional simplifications which are necessary before objective evaluation via measurement can be employed. But they will also introduce the complications associated with viewing populations as combinations of age groups, together with the spatial component, so that population can be seen to grow over time in a spatially differentiated fashion. To remain for the moment with the 'how' and 'why' question in the time dimension we will turn to the works of Malthus and Marx and to the influential and opposing theories and 'laws' which they propounded by way of answers.

7

Socio-Economic Status of Slum Dwellers

The study region is municipal area situated in dist Pratapgarh, which sites in Gangetic plain in UP. It spreads over 12 km2 with total population of 71999 persons. In the age of urbanization in developing countries like India, the slums are like white spot on the body which is caused by uneven development. To sort out the urban problems, socioeconomic study of the city is necessary and if we want to have clean cities in near future, the studies of slums are not only important but demand of time also. Therefore a sample area has been randomly taken for the socioeconomic study. The parameters for demarcation slums are water supply condition, sanitation and road, employment, housing condition, sufficient living area.

To sum up, the study reveals that a particular caste resides in a particular slum and these castes are related to either Pal or Harijan and Muslims. The slums lack with water, sanitation, road facility and it also denotes that the slum dwellers are unaware of programs launched by the Government. Pratapgarh District, one of the oldest Districts of UP, which came into existence in the year 1858, is situated in eastern part of UP. According to JNU's development list based on socio-economic-agricultural, Pratapgarh stands in last category. So it is an undeveloped district. Pratapgarh district famous for its Aonla and people gives alias "AONLA NAGRI".

The district, which forms from a part of Faizabad division, is named after it headquarter town Bela Pratapgarh commonly known as Pratapgarh. When district was constituted in 1858 it's headquarter established at Bela which come to know as Bela Pratapgarh. The name Bela presumably being derived from the temple of "Bela Bhawani", sites on the bank of river SAI. The study region districts headquarter; Bela Pratapgarh is situated at Allahabad-Faizabad national highway no. 96, at a distance of 39 km from Sultanpur and 60 km from Allahabad. Bela Pratapgarh city is a secondary town and spreads over 12 km2 with total population of 71999 persons (as per 2001 census). It is municipal area, which is divided in 25 wards so far. In total population of Bela Pratapgarh, percentage of OBC population is 31.58% and SC is 7.97%. The percentage of male and female population is 52.67% and 47.32%. The SEX ratio of Bela Pratapgarh is 898 which are quite lower than the district ratio (1004) and literacy rate is 71.039% which is higher than the district literacy ratio (58.67 percent). In Bela Pratapgarh, male literacy rate 77.03 percent followed by female 64.36 percent and district male literacy rate 74.61 percent followed by female literacy rate 42.63 percent.

The Study Area:-Study area of slums spread in all over the city in small patches. In 1997 District Urban Development Agency (DUDA) was stabilized and DUDA has declared 3 slum areas, these registered slums are Patkohli, Padawa and Karanpur. But there are some unregistered slum areas like Azadnagar, Khuskhuswapur, Pitaikapurwa, Bholiapur etc. Among these slum areas Patkoli ward is largest slum area with the 160 households. In Patkoli ward slum area is found in 3 patches-Ziriyamau, Patkoli and Bela Ghat; Pitaikapurwa with 2 patches-pitai-ka-purwa and naibasti. Other slum areas have 30 to 60 households.

Ward-name Basti name No of households

- Patkoli ziriyamau, patkoli, bela gaht 160
- Aspatal ward Azadnagar 60
- Bali Pur Khuskhuswapur 35
- Pitai ka Purwa Naibasit, pure pitai 60
- Padawa harizen basti 30
- Karanpur Karanpur 35
- Bholiapur Bholiapur 25.

Objective of the Study:-Every citizen has the right to live in a good living condition with employment, safety and other facilities. If any area left undeveloped, creates crisis in the mode of crime, unsocial activities, uneducated and unaware persons, low-level living condition, inadequate housing condition, unmoral citizens and unhealthy children with the unhealthy future of city and country. Now governments aware of the slum Basties of cities and launched many programmes to improve the condition of slum dwellers and his proficiency. But the game does not over, commonly seen that unregistered slums site over Urban Fringe area or at inner border of city, because of special spatial condition policies never imposed its effect in full fledged from. In city area, which is defined by the municipality area, plans take its own time to produce its impacts. Since, slums are present before the executions of development plans; therefore many problems stand in front of city and the citizens in form of crime, violence, poverty etc. These causes also exist in the study area. The objective behind the study aimed at to evaluate the socioeconomic condition and to access the impact of government programmes upon slum dwellers of Pratapgarh City (Bela).

Methodology:-Every city has two types of slum areas i.e. registered and unregistered. The parameters for demarcating the slum are water supply condition, sanitation and road, housing condition. After that, to study the socioeconomic status, a random sampling method has been adopted. For the survey and sample collection, a questionnaire has been prepared and door to door survey has been done. Because, Patakoli ward has large number of households, so, in this area 20 households were randomly selected for the study and in other areas 8 households have randomly taken. To evaluate the condition of dwellers, center tendency method is used.

Social Status:-In Bela Pratapgarh, slum dwellers are Hindus (64.44 percent) and Muslims (35.64 percent). In the total population of slum dwellers OBC with 57.14 percent, are in majority followed by SC with 33.65 percent and General with 9.2 percent.

Religion	***Category Percentage***
Hindu General	2.955
OBC	47.783
SC	49.261
Muslim General	20.595
OBC	74.107
SC	5.357

Among Hindus, SC population is 49.261 percent and followed by OBC with 47.783 percent and General with 2.955 percent. In Muslims, OBC population is in majority with 74.107 percent followed by General 20.595 percent and SC with 5.357 percent. In slum areas sex ratio is 944 is greater than the Bela Pratapgarh sex ratio (898) and State sex ratio (898) but lower than district sex ratio (1004). In slum areas Hindus sex ratio (1071) is higher than Muslim sex ratio (778).

Literacy:-The overall literacy rate in slum areas is 56 percent with male and female literacy 65 percent and 46 percent against 71.039 percent in Bela Pratapgarh with male and female literacy 77.03 percent and 64.36 percent. Literacy rate of slum areas, near to district rate (58.67 percent) and UP state rate (57.36 percent) but it lower than the National rate (64.8 percent).

The male literacy rate of slum areas 65 percent is quite lower to district rate (74.61 percent), national rate (75.8 percent) and state rate (70.23 percent) but female literacy rate 46 percent is near to district rate (42.63 percent), state rate (42.98 percent) and lowers than the national ratio (53.7 percent).

In slum areas Primary educated persons are in majority with 60.71 percent followed by Madhyamic (28.27 percent), Inter (5.089 percent), Graduate (2.68 percent) and Post-Graduate (1.071 percent). The vocational trained persons are 1.09 percent.

Income:-Total working people percentage of slum dwellers is near about 24.14 percent. Among these persons, 60 percent people do not know the nature of day's work.

The average of working day in a month stands between 16 to 20 days. Over all average income of slum dwellers is quite lower with 13.03 Rs a day/person against the urban poverty line, which is near about 17 Rs. Average income of a working people 55.98 Rs for doing work 7 hour and 36 min in a day, which is quite lower than the UP Government regulation.

Housing condition:-In slum areas housing conditions are found in a very poor position. In sample area 46.34 percent of Kchcha houses made by mud, plastic and bamboo, 21.95 percent of Pakkaa houses made by bricks and 31.71 percent houses made by mud and bricks which have greater part Kachcha and smaller part Pakka. So, 78 percent of households face problem of housing.

Sanitation, Water, Road and Health:-In slum area sanitation and road facility are totally unavailable, for example except karanpur, ziriyamau, ptakholi, bela ghat, padava have Kaccha road and azad nagar, khuskuswapur, naibasati and pitai ka purwa have Kharanja road. In these areas drinking water facilities totally unavailable, except Azadnagar, in other ward water source is well or government hand pump, at per 25 households. Also in Azadnagar drinking water facilities totally unavailable and dwellers get drinking water from hand-pumps of rich neighbours. In other areas, if there is water supply, its condition is very poor. Generally water supply pipes are sunk in drain water. Hence, the water can not be used being coal black. In slum area latrine facilities are in poor condition, only 20% households have own latrines and other 80% uses the open lands. The wastages of households dump by householder near the open land or the path. As far as, the total environment of slums does not good, in all these areas children are suffering from viral infections, boils, fever etc. for ever.

Social Activities:-In crime, violence and immorality slum areas are leading portion of the city. In study area violence seems as quarrel, fight or abuse in home with wife, children, parents etc. or with the neighbours. The day life of the dwellers start and end with the quarrel and it happens because of drinking liquor. Immorality can be seen in the form of gambling. A sentence told by an old man to me in Srirama about the gambling, pinched me a lot. He told, "He bhaiya agar in nanha-nanha larikan ka juaa chudawai detya tau enkar kalyan hoi jaatai." (If you can escape younger Boys from gambling, then they would lead a good life). In these areas some families are vagabonds. Police keeps an eye over them, arrest and warn. Because most of them many times involved in unsocial and immoral activities like snatching, thieving, gambling etc.

Government Plans and Awareness:-District Urban Development Agency is the coordinating agency at district level for formulating different developmental schemes in urban areas. The following schemes/ programmes are being implemented under the supervision of DUDA.

Swarna Jayanti Sahari Rojgar Yojana, which is currently in operation with effect from 1.12.97, is a substitution of the earlier programmes like Urban Basic Services for the Poor (UBSP), Prime Minister Integrated Urban Poverty Eradication Yojana (NRY), Environmental Improvement of Urban Slums (EIUs). Under the scheme SJSRY, the following programmes are being implemented in all the ULBs of this district.

- Community structure (Information Education and Communication (IEC) Component.
- Urban Self Employment Programme (USEP) (subsidy)
- Development of Women and Children in Urban Areas (DWCUA)
- Thrift and Credit Society
- Urban Wage Employment Programme (UWEP)

- U.S.E.P (Skill up gradation Training)
- Administration and Office expenses (A & O.E.)

About these programmes and DUDA slum dwellers unaware and in sample area only one person know about DUDA and its U.S.E.P. programme. It shows the lack of publicity of programmes and unawareness of slum dwellers. Some advantages they gain from common programmes like Antyodaya yojana, Mid-day-Meal programme and Scholar-Ship programmes.

To conclude; we can say that a big number of people, known as slum dweller, are forced to live below poverty line. The main reason of their poverty is illiteracy and lack of will power to improve his condition. They always look for the government helps; debt, relief etc. and wait for them, this type of mentality and government plans make them paralysed instead of self reliance. They lack behind good education, health, sanitation and economy. There are Governments plans to improve their poor condition. But these programmes are executed half heartedly. There is need to focus their problems and a strong heart and mind to execute these programmes. If it happens so, the day will come with hope for them and for the people who are affected by these slum dwellers.

Infrastructure Financing and Emerging Pattern of Urbanisation: A Perspective

Large sections of planners and policy makers in the country have argued that there exists no serious problem of infrastructural deficiency that can not be tackled through management solutions. All that is needed is to restructure the system of governance, legal and administrative framework in a manner that the standard reform measures can be implemented. Reduction of public sector intervention, ensuring appropriate prices for infrastructure and civic amenities through elimination or reduction of subsidies, development of capital market for resource mobilisation, facilitating private and joint sector projects, simplification of legislative system to bring about appropriate land use changes and location of economic activities etc. are being advocated as the remedial package.

The public sector and other para-statal agencies that had been assigned the responsibility of producing and distributing infrastructural facilities have come in for sharp criticism on grounds of inefficiency, lack of cost effectiveness, resulting in continued dependence on grants for sustenance. Some kind of "financial discipline" has already been imposed by the government and Reserve Bank of India, forcing these agencies to generate resources internally and borrow from development cum banking institutions, and, in a few cases, from capital market at a fairly high interest rate. This has restricted their areas of functioning and, what is more important, changed the thrust of activities.

Solutions are being found also in terms of their efficient, transparent and decentralised management of the facilities. With the passing of the 74th Amendment to Indian Constitution and corresponding legislations, amendments, ordinances etc. at the state level, decentralisation has become the keyword in governance. The vacuum created by the limited withdrawal of the state in the provision of infrastructure is sought to be filled up also through non-governmental organisations (NGOs) and community based organisations (CBOs), besides the local authorities. The enthusiasm for the above package of "management solutions", both among the international as also national organisations, is responsible for the issues concerning their impact on settlement structure and access of the poor to the infrastructural amenities not receiving adequate attention among researchers. However, given the disparity in economic strength of the towns and cities and their unequal access to capital market and public institutions, this perspective would enable the larger cities to corner much of the advantage from the system.

Also, large sections of urban poor are likely to be priced out of the formal systems of service delivery. A few researchers have pointed out that the indifference on the part of policy makers on these issues would institutionalise inequality in infrastructural facilities and accentuate disparity in the levels of economic development. In the light of the dominant perspective and criticisms thereof, as discussed above, the present paper overviews the recent initiatives for resource mobilisation for infrastructural investment and analyses their impact on the structure of settlements, access of poor to infrastructural facilities and process of segmentation within the cities. The analysis has been done by taking into consideration the recent changes in labour and capital market and land management\ development practices.

The second section in the paper analyses the shift from budgetary support to institutional finance for infrastructure development and discusses the new arrangements for mobilising resources, in the wake of the strategy of economic liberalisation and the changes in urban governance as a consequence of the 74th Constitutional Amendment. Assigning contracts to private agencies for providing infrastructural services has also been examined critically. Furthermore, infrastructure development\ improvement projects being implemented by NGOs and local communities within a participatory framework have also been evaluated. The third section overviews the scenario of infrastructure financing likely to emerge in the next couple of decades and its impact on the deficiency of amenities across states and size classes. In the last section, an attempt has been made to put forward a perspective for infrastructural development that can bring about a balanced regional development in the country and fulfill other socioeconomic objectives to which the country stands committed through the Constitution and Directive Principles of State Policy. It also outlines a strategy for intervention that can help in attaining the objectives.

Financing Infrastructure Development: Recent Trends and Institutional Initiatives

Budgetary Support and Institutional Borrowings

The system of managing and financing infrastructural facilities has been changing significantly since the mid-eighties. The Eighth Plan (1992-97) envisaged cost recovery to be built into the financing system. This has further been reinforced during the Ninth Plan period (1997-2002) with a substantial reduction in budgetary allocations for infrastructure development. A strong case has been made for making the public agencies accountable and financially viable. Most of the infrastructure projects are to be undertaken through institutional finance rather than budgetary support. The state level organisations responsible for providing infrastructural services, metropolitan and other urban development agencies are expected to make capital investments on their own, besides covering the operational costs for their infrastructural services. The costs of borrowing have gone up significantly for all these agencies over the years. This has come in their way of their taking up schemes that are socially desirable schemes but are financially unremunerative. Projects for the provision of water, sewerage and sanitation facilities etc., that generally have a long gestation period and require a substantial component of subsidy, have, thus, received a low priority in this changed policy perspective.

Housing and Urban Development Corporation (HUDCO), set up in the sixties by the Government of India to support urban development schemes, had tried to give an impetus to infrastructural projects by opening a special window in the late eighties. Availability of loans from this window, generally at less than the market rate, was expected to make state and city level agencies, including the municipalities, borrow from HUDCO. This was more so for projects in cities and towns with less than a million population since their capacity to draw upon internal resources was limited.

HUDCO finances even now up to 70 per cent of the costs in case of public utility projects and social infrastructure. For economic and commercial infrastructure, the share ranges from 50 per cent for the private agencies to 80 per cent for public agencies. The loan is to be repaid in quarterly installments within a period of 10 to 15 years, except for the private agencies for whom the repayment period is shorter. The interest rates for the borrowings from HUDCO vary from 15 per cent for utility infrastructure of the public agencies to 19.5 per cent for commercial infrastructure of the private sector. The range is much less than what used to be at the time of opening the infrastructure window by HUDCO. This increase in the average rate of interest and reduction in the range is because its average cost of borrowing has gone up from about 7 per cent to 14 per cent during the last two and a half decade.

Importantly, HUDCO loans were available for upgrading and improving the basic services in slums at a rate lower than the normal schemes in the early nineties. These were much cheaper than under similar schemes of the World Bank. However, such loans are no longer available.

Also, earlier the Corporation was charging differential interest rates from local bodies in towns and cities depending upon their population size. For urban centres with less than half a million population, the rate was 14.5 per cent; for cities with population between half to one million, it was 17 per cent; and for million plus cities, it was 18 per cent. No special concessional rate was, however, charged for the towns with less than a hundred or fifty thousand population that are in dire need of infrastructural improvement, as discussed above. It is unfortunate, however, that even this small bias in favour of smaller cities has now been given up. Further, HUDCO was financing up to 90 per cent of the project cost in case of infrastructural schemes for "economically weaker sections" which, too, has been discontinued in recent years. HUDCO was and continues to be the premiere financial institution for disbursing loans under the Integrated Low Cost Sanitation Scheme of the government. The loans as well as the subsidy components for different beneficiary categories under the scheme are released through the Corporation. The amount of funds available through this channel has gone down drastically in the nineties.

Given the stoppage of equity support from the government, increased cost of resource mobilisation, and pressure from international agencies to make infrastructural financing commercially viable, HUDCO has responded by increasing the average rate of interest and bringing down the amounts advanced to the social sectors. Most significantly, there has been a reduction in the interest rate differentiation, designed for achieving social equity.

An analysis of infrastructural finances disbursed through HUDCO shows that the development authorities and municipal corporations that exist only in larger urban centres operate have received more than half of the total amount. The agencies like Water Supply and Sewerage Boards and Housing Boards, that have the entire state within their jurisdiction, on the other hand, have received altogether less than one third of the total loans. Municipalities with less than a hundred thousand population or local agencies with weak economic base often find it difficult to approach HUDCO for loans. This is so even under the central government schemes like the Integrated Development of Small and Medium Towns, routed through HUDCO, that carry a subsidy component.

These towns are generally not in a position to obtain state government's guarantee due to their uncertain financial position. The central government and the Reserve Bank of India have proposed restrictions on many of the states for giving guarantees to local bodies and para-statal agencies, in an attempt to ensure fiscal discipline. Also, the states are being persuaded to register a fixed percentage of the amount guaranteed by them as a liability in their accounting system. More importantly, in most of the states, only the para-statal agencies and municipal corporations have been given state guarantee with the total exclusion of smaller municipal bodies. Understandably, getting bank guarantee is even more difficult, specially, for the urban centres in less developed states and all small and medium towns.

The Infrastructure Leasing and Financial Services (ILFS), established in 1989, is coming up as an important financial institution in recent years. It is a private sector financial intermediary wherein the Government of India owns a small equity share. Its activities have more or less remained confined to development of industrial-townships, roads and highways where risks are comparatively less. It basically undertakes project feasibility studies and provides a variety of financial as well as engineering services. Its role, therefore, is that of a merchant banker rather than of a mere loan provider so far as infrastructure financing is considered and its share in the total infrastructural finance in the country remains limited. ILFS has helped local bodies, para-statal agencies and private organisations in preparing feasibility reports of commercially viable projects, detailing out the pricing and cost recovery mechanisms and establishing joint venture companies called Special Purpose Vehicles (SPV). Further, it has become equity holders in these companies along with other public and private agencies, including the operator of the BOT project (Mathur 1999). The role of ILFS may, thus, be seen as a promoter of a new perspective of development and a participatory arrangement for project financing. It is trying to acquire the dominant position for the purpose of influencing the composition of infrastructural projects and the system of their financing in the country.

Mention must be made here of the Financial Institutions Reform and Expansion (FIRE) Programme, launched under the auspices of the USAID. Its basic objective is to enhance resource availability for commercially viable infrastructure projects through the development of domestic debt market. Fifty per cent of the project cost is financed from the funds raised in US capital market under Housing Guaranty fund. This has been made available for a long period of thirty years at an interest rate of 6 percent, thanks to the guarantee from the US-Congress. The risk involved in the exchange rate fluctuation due to the long period of capital borrowing is being mitigated by a swapping arrangement through the Grigsby Bradford and Company and Government Finance Officers' Association for which they would charge an interest rate of 6 to 7 percent. The interest rate for the funds from US market, thus, does not work out as much cheaper than that raised internally.

The funds under the programme are being channelled through ILFS and HUDCO who are expected to raise a matching contribution for the project from the domestic debt market. A long list of agenda for policy reform pertaining to urban governance, land management, pricing of services etc. have been proposed for the two participating institutions. For providing loans under the programme, the two agencies are supposed to cxamine the financial viability or bankability of the projects.

This, it is hoped, would ensure financial discipline on the part of the borrowing agencies like private and public companies, municipal bodies, para-statal agencies etc. as also the state governments that have to stand guarantee to the projects. The major question, here, however is whether funds from these agencies would be available for social sectors schemes that have a long gestation period and low commercial viability. Institutional funds are available also under Employees State Insurance Scheme and Employer's Provident Fund.

These have a longer maturity period and are, thus, more suited for infrastructure financing. There are, however, regulations requiring the investment to be channelled in government securities and other debt instruments in a "socially desirable" manner. Government, however, is seriously considering proposals to relax these stipulations so that the funds can be made available for earning higher returns, as per the principle of commercial profitability.

There are several international actors that are active in the infrastructure sector like the Governments of United Kingdom (through Department for International Development), Australia, Netherlands. These have taken up projects pertaining to provision of infrastructure and basic amenities under their bilateral co-operation programmes. Their financial support, although very small in comparison with that coming from other agencies discussed below, has generally gone into projects that are unlikely to be picked up by private sector and may have problems of cost recovery. World Bank, Asian Development Bank, OECF (Japan), on the other hand, are the agencies that have financed infrastructure projects that are commercially viable and have the potential of being replicated on a large scale. The share of these agencies in the total funds into infrastructure sector is substantial. The problem, here, however, is that the funds have generally been made available when the borrowing agencies are able to involve private entrepreneurs in the project or mobilise certain stipulated amount from the capital market. This has proved to be a major bottleneck in the launching of a large number of projects. Several social sector projects have failed at different stages of formulation or implementation due to their long pay back period and uncertain profit potential. These projects also face serious difficulties in meeting the conditionalities laid down by the international agencies.

Borrowings by State Government and Public Undertakings from Capital Market

A strong plea has been made for mobilising resources from the capital market for infrastructural investment. Unfortunately, there are not many projects in the country that have been perceived as commercially viable, for which funds can easily be lifted from the market. The weak financial position and revenue sources of the state undertakings in this sector make this even more difficult. As a consequence, innovative credit instruments have been designed to enable the local bodies tap the capital market.

Bonds, for example, are being issued through institutional arrangements in such a manner that the borrowing agency is required to pledge or escrow certain buoyant sources of revenue for debt servicing. This is a mechanism by which the debt repayment obligations are given utmost priority and kept independent of the overall financial position of the borrowing agency. It ensures that a trustee would monitor the debt servicing and that the borrowing agency would not have access to the pledged resources until the loan is repaid.

The most important development in the context of investment in infrastructure and amenities is the emergence of credit rating institutions in the country. With the financial markets becoming global and competitive and the borrowers' base increasingly diversified, investors and regulators prefer to rely on the opinion of these institutions for their decisions. The rating of the debt instruments of the corporate bodies, financial agencies and banks are currently being done by the institutions like Information and Credit Rating Agency of India (ICRA), Credit Analysis and Research (CARE) and Credit Rating Information Services of India Limited (CRISIL) etc. The rating of the urban local bodies has, however, been done so far by only CRISIL, that too only since 1995-96.

Given the controls of the state government on the borrowing agencies, it is not easy for any institution to assess the "functioning and managerial capabilities" of these agencies in any meaningful manner so as to give a precise rating. Furthermore, the "present financial position" of an agency in no way reflects its strength or managerial efficiency. There could be several reasons for the revenue income, expenditure and budgetary surplus to be high other than its administrative efficiency. Large sums being received as grants or as remuneration for providing certain services could explain that. The surplus in the current or capital account can not be a basis for cross-sectional or temporal comparison since the user charges permitted by the state governments may vary.

More important than obtaining the relevant information, there is the problem of choosing a development perspective. The rating institutions would have difficulties in deciding whether to go by measures of financial performance like total revenue including grants or build appropriate indicators to reflect managerial efficiency. One can possibly justify the former on the ground that for debt servicing, what one needs is high income, irrespective of its source or managerial efficiency. This would, however, imply taking a very short-term view of the situation. Instead, if the rating agency considers level of managerial efficiency, structure of governance or economic strength in long-term context, it would be able to support the projects that may have debt repayment problems in the short run but would succeed in the long run.

The indicators that it may then consider would pertain to the provisions in state legislation regarding decentralisation, stability of the government in the city and the state, per capita income of the population, level of industrial and commercial activity etc. All these have a direct bearing on the prospect of increasing user charges in the long run. The body, for example, would be able to generate higher revenues through periodic revision of user-charges, if per capita income levels of its residents are high.

The rating agencies have, indeed, taken a medium or long-term view, as may be noted from the Rating Reports of various public undertakings in the recent past. These have generally based their rating on a host of quantitative and qualitative factors, including those pertaining to the policy perspective at the state or local level and not simply a few measurable indicators. The only problem is that it has neither detailed out all these factors nor specified the procedures by which the qualitative dimensions have been brought within the credit rating framework, without much ambiguity.

Political Decentralisation and Investment in Infrastructure and Basic Services by Local Bodies

The resource scarcity being faced by the central and state governments and the consequent reduction of current and capital expenditures on infrastructure and social sectors in recent years have created serious uncertainty with regard to provision of basic amenities, particularly to the poor. Earlier, the role of central and state governments in local affairs was not clearly defined. It consisted of ad-hoc and fragmented efforts at programmatic level. Since mid eighties, however, a process of shifting the responsibility to the local level has manifested clearly. Political decentralisation through Constitutional Amendment Act has been hailed as a panacea for the problems of infrastructural deficiency in urban centres.

It is argued that the Act enables the local bodies to undertake planning and development responsibility as also mobilise resources for infrastructural investment. Transferring of responsibilities to local bodies, without examining their economic base and resource raising capacity or making provision of adequate transfer of funds, may, however, have serious consequences. Several studies suggest that there exists significant disparity in the income levels of local bodies and its various components across size class of urban centres. A study by the National Institute of Public Finance and Policy (NIPFP 1995) reveals that the per capita (own) revenue for D Class cities, having population above five hundred thousand, was more than three and a half times that of A class towns, having population below one hundred thousand, in the early nineties. The tax and non-tax revenue together constituted 90 per cent of the total revenue in case of the former while the figure for the latter was 70 per cent only. Correspondingly, the percentage share of grants in total revenue for the D class cities was only 5 per cent while that for the lower class towns was as high as 18 per cent. The high dependence on external grants for the smaller towns would be a major handicap in their undertaking the development responsibility on their own. One may, thus, argue that the larger cities are financially stronger and can take up public works and social infrastructure projects on their own which is not so for smaller towns.

The NIPFP study unfortunately does not give further break-ups for the towns falling in A class by considering the size classification followed by the Population Census. Also, the sample does not include many towns with less than 50,000 population and none with less than 10,000. When these towns are included in the sample, the disparity in revenue earnings across size class of urban centres would work out as much sharper (National Institute of Urban Affairs 1983). A large majority of these towns have come to depend increasingly on grants-in-aid, primarily due to their poor economic base and incapacity to mobilise adequate tax and non-tax revenues. With the decline in central or state assistance in recent years, it is not surprising that most of these towns do not make any investment for improving infrastructure and basic services. This has compounded their problems of inadequacy of basic amenities.

People in small and medium towns in the country, particularly those with less than 50,000 population, have low per capita income due to lack of employment opportunities in the organised sector, low incidence of other secondary activities, poverty induced growth of tertiary employment etc. Instability in their economy is reflected in high fluctuation\ variation in their demographic growth over time as also across regions, as noted above. Understandably, many among the small and medium towns are not in a position to invest funds for development of civic amenities. The Constitutional Amendment, making the civic bodies increasingly dependent on their own tax and non-tax resources, would further increase the disparity in the level of services and economic infrastructure across size class of urban centres. This would adversely affect the level of basic services in these towns and their capacity to absorb future growth of population or attract new economic activities.

In an era, wherein obtaining funds for investment from private sector or capital market has become the critical factor, it is not surprising that the exercises of preparing Master Plans have been thoroughly discredited. Indeed, the city level Master Plans in recent years have been referred to more for violation than for compliance since no fund is available for their implementation. Given this emerging scenario, the option of going for project preparation in formal or informal consultation with the stockholders through the intermediation of the financial institutions, seems to be an easy way out. This, the local governments in large cities would find hard to resist. It must, however, be pointed out that preparing a Master plan to answer the needs and aspirations of local people (in different social and income brackets) and then breaking it down to a set of meaningful projects is a different exercise from that of identifying projects with high credit rating that would attract corporate investment. Assigning the responsibility of development planning to local bodies, thus, does not automatically enable them to design projects, based on a macro perspective for the town or region. The data available at the micro level are not adequate to support the optimism that the Amendment would indeed enhance the resources, particularly of the local bodies and enhance their capacity to invest in infrastructure.

Obtaining a hefty loan from capital market or even an international organisation may lead to dis-empowerment of local bodies. Sanctioning of the loan has been subject to the latter maintaining budgets of the sectors receiving the funds separate from the general budget. Bombay Municipal Corporation, for example, was obliged to maintain water supply and sewerage budget on accrual basis while the other budgets of the corporation are maintained on cash basis. This was the pre-condition for getting the World Bank loan under the International Development Assistance credit scheme. Similarly, a study on Pune Municipal Corporation under the FIRE project, proposes a budgeting discipline for the local bodies such that they maintain separate accounts for not merely capital and current expenditures but also for different set of sectors (Mehta and Satyanarayana (1996). Proposals have been put forward for revising the user charges along with the norms for sectoral transfers such that a group of sectors can become independent of the general municipal budget. This is considered necessary for attracting capital investments in future years.

Land as a Source of Infrastructure Financing

Many of the state governments are trying to create a few "global centres of the future" while the city governments are engaged in competition with their counterparts, within the state or country, to grow up to that stature by attracting private corporate investment. The prospective global cities, however, face a major problem in attracting local or international investment in infrastructural facilities. The problem is that of scarcity of land within the central city and other prime locations wherein such facilities are in high demand. An ingenious method has been worked out to solve the problem. The agencies like World Bank, USAID etc. have recommended that the Floor Space Index (FSI) in the central areas of the city should be increased so that multi-storied structures can come up, providing space for business houses, commercial activities and high income residential units. The policy of giving permission for vertical growth at a high price or selling of extra FSI in central and business districts has been welcomed by many local bodies also as an easy way of generating resources for infrastructural development. There is further incentive to this since sanctioning of loans by the international agencies is often contingent on the acceptance of a higher FSI in the central city. This has led to creation of a few high-density business and high-income residential districts and pushing out of the households that can not afford the costs.

Land is provided for infrastructural facilities and upcoming industrial and commercial houses through government intervention as also by activating the land market. Steps are taken to facilitate changes in land-use pattern through simplification of legal and administrative procedures and enabling the market to push "low valued" activities out of the city core. The low income and slum colonies are the obvious candidates for relocation in city peripheries. The shift is carried out by the state or local governments, often directly through eviction of slum dwellers, hawkers, pavement dwellers etc. Sometimes, it is done indirectly and discreetly through slum improvement schemes, "rehabilitating" them outside the city limits. Unfortunately, that has been done mostly without making any provision for alternate employment opportunities for the displaced workers. This has led to high disparity in population density, quality of life and segmentation of the cities into rich and poor colonies.

Management Contracts for Provision of Services by Private Agencies

Widening gaps between demand and supply of infrastructural services have necessitated involvement of private sector and designing of alternate institutional arrangements. Different forms of participation have been worked out for investment and management of the services, with varying levels of responsibility and cost sharing between private and public agencies. The most commonly used form is contracting out the tasks of capacity creation, production, distribution and management of one or a set of services to private companies for certain period. It has been argued that this would result in substantial cost saving and, at the same time, make it possible for the companies to conduct their business with profit.

In order to ensure that the contractual agreements with private or joint sector companies are executed, supervised or monitored under strict regulatory control, new regulatory\ supervisory arrangements are being proposed. Under the old system of state control, the public agencies had failed in meeting the social objectives and catering to the needs of the poor. Designing and implementing the regulatory controls in a liberalised regime is, therefore, far more challenging, in case we are serious about these objectives. Importantly, UNCHS has launched an urban management programme (UMP), with offices in many developing countries including India, for creating an enabling environment for such participatory management practices and assist the local bodies in drawing up terms and conditions of the contracts, involving community participation in management. Besides, it facilitates the local bodies in establishing contacts with possible participating companies and often conducts negotiations on their behalf. These moves in urban areas have had some success as a number of cities have come forward to institutionalise such contractual arrangements. It may, however, be pointed out that a number of companies being engaged in the provision of the services and their making profits or local bodies reducing their financial burden through privatisation of services are by themselves no indication of success of the arrangement. Such a judgement is possible only after the issues concerning quality of the services and, more importantly, the coverage of the urban poor have been analysed in a rigorous manner.

It is difficult to empirically determine the benefits and costs of this privatisation move, as it is a recent phenomenon. Reliable data on revenue and expenditure of local bodies before and after the sub-contracting arrangement or the profitability of the private agencies are yet not available. The critical information in this context would, however, be with regard to the quality of the services and the coverage of the vulnerable sections of the city population, as noted above. Unfortunately, there exists no database covering these aspects at national, state or local level. These would have to be obtained through field survey at the micro level. It is nonetheless very surprising that none of the studies conducted in connection with privatisation of services or subcontracting arrangements has attempted to focus on these aspects seriously.

An overview of studies on the contractual agreements through their progress reports and other official documents reveals that there have been problems even in operationalising the working arrangements and norms for day to day functioning even at the stage of capital investment. This is largely because India does not have much experience of partnership arrangements in the field of urban infrastructure. Many have got trapped in serious difficulties after the formal launching of the projects. In Vadodara, for example, the Municipal Corporation had worked out an arrangement with a private company to make pellets from waste, to be used as fuel. Unfortunately, the company faced serious difficulties in establishing working arrangements with the local bodies and had to finally close down its unit. Similar difficulties have been encountered in a number of other projects.

Solid waste management is one service where the local bodies have shown interest in involving the private sector. The method of collection often involves participation of citizens' groups and non-governmental organisations (NGOs). Many groups have shown willingness to spare time or money for an improved system of garbage disposal, which is behind the success stories coming from a number of cities. There is, however, no reason to believe that the coverage of the poor areas or households through basic amenities would increase after a private company takes over the responsibility.

There have been demands from the companies to hike the user charges commensurate with inflation and even more. But the low income and slum colonies have resisted such moves, resulting in their exclusion from the private sector based delivery system. In such a situation, the people in these localities have not been able to raise their voices or complain against the municipal body as the latter would take the plea of passing on the responsibility to an outside agency. This has led to fragmentation of responsibility of service delivery, with marginalised areas\ people remaining under serviced or unserviced, as noted above. As a consequence, there has been serious health risk for the entire urban community that may result in breaking of epidemics from time to time.

Involving Community in Infrastructure Development

Community participation in urban development projects has a history dating back to the early seventies when Urban Basic Service Programme was launched by the central government with assistance from UNICEF. The projects covered only a few slums in select number of cities. There was no perspective for building a network to cover all the slums even within the selected cities and, consequently, the solutions pursued were local in nature. Attempts were made to do social mobilisation by creating community groups and involving these in implementation of the project at the grassroots level. Community was, thus, viewed merely as an agent, providing support to state sponsored development activities to ensure better implementation, in the initial experiments of community participation.

In subsequent years it was realised that community involvement not only results in effective implementation of the projects but also leads to better designing and substantial reduction in operational costs. Following this, community was often involved not just for project implementation and supervision of work but also in designing the project. All these initiatives notwithstanding, community participation remained a state sponsored activity until the late seventies.

Community participation as a component of development strategy gained currency in the eighties as a consequence of failure of public agencies and growing deficiency in the level of basic amenities. It was argued that the community can help not only in social mobilisation but also in raising financial resources that the local authorities need very badly. In many cases, it became possible to have substantial reduction in project cost as the prospective beneficiaries provided their labour free or at a wage rate much below that in the market. Further, the pressure of peer group under participatory arrangement resulted in better monitoring, more productive engagement of the beneficiaries in the project and better recovery of development loans sanctioned to individuals. The community was mobilised not merely for making contributions in terms of ideas and labour but also for sharing a part of the capital and current expenditure. Only, there was demand for making credit available outside the formal institutional structure, which was beyond the access of the slum communities. In several cases, attempts were made to build mechanisms at community level to ensure timely repayment of loans. The most innovative form of community participation in infrastructural projects, which has been hailed as an major achievement in the nineties, is the neighbourhood and slum networking schemes, launched with substantial financial support from the state or central government. The people in the project area have been associated with all stages in the development of the project and are given the overall responsibility of monitoring it as also maintaining the services. The most celebrated case of community involvement in the nineties is the Slum Networking Project (SNP) Indore. The second case-still in the process of implementation-is SNP Ahmedabad project. Similar projects are being tried out in a number of cities in Maharashtra by involving the local community and the NGOs. These projects often involve provision of roads, pavements, water supply, sewerage, drainage and waste disposal facilities, street lighting, water taps and toilets connected to a network provided on individual or community basis. Sometimes projects are designed to take up not the entire agenda of slum improvement but provision of one or a few of the infrastructural facilities. It is argued that "the slums typically cover only about 5 per cent of the land area of cities... it is thus possible to have a massive impact on the entire city and its infrastructure by working in these very small areas. Concentrating resources in these neediest areas would, thus, work out as highly cost effective". This approach, understandably, helps in mainstreaming the issue by underlining the fact that investment in slum infrastructure is a critical component of urban development, which can yield a high rate of economic and social return.

The advocates of the projects have argued that private industries can be motivated to join these projects as they have an interest in the overall hygiene and health of the city because that determines their business prospects in the long run. They can, therefore, be persuaded to share the cost of slum development. This perspective has allowed the advocates of the participatory approach to place the slum-linked activities within the mainstream of urban development strategy.

The Emerging System for Financing Infrastructure Development and Its Impact: A Vision for 2020

It would be interesting to speculate on the nature of the system of financing for infrastructural sector in about a couple of decades from now and what would be its impact on the availability of infrastructural facilities across states and size class of settlements. More importantly, it would be worthwhile to assess the possible impact of this new system on the urban structure in the country.

Understandably, the budgetary support for infrastructural projects, coming from central or state governments would be very low in two decades from now. As a consequence, the investment in small and medium towns, particularly in the less developed states would be very small. The projects for the provision of basic amenities in slums or low income areas, even in large cities not having clear stipulations for total cost recovery within a reasonable period-unlikely to be fancied by public agencies. Unfortunately, investment in infrastructure in small towns in backward statea undertaken with finances from organisdations like HUDCO, ILFS will be very low. The would not find favour even under the institutional arrangements of USAID FIRE project nor of private or joint sector company because of uncertainties with regard to their commercial viability.

On the face of the difficulties in obtaining loans from financial institutions, efforts are on to develop the capital market so that the public utilities and other para-statal agencies can mobilise resources directly by issuing bonds and other credit instruments. This strategy, being operationalised with support from financial intermediaries, including credit rating institutions, has serious spatial implications. Given the need to keep user charges for the basic facilities within certain limits, many of the public agencies may not be able to raise funds from the market and launch major expansion in activities in the next few decades. Further, the cities issuing bonds, SDOs etc. for undertaking investment in infrastructure, have been forced to escrow or pledge their regular earnings from octroi, property tax, grants from the state etc. as a guarantee for debt servicing. An analysis of the arrangements, worked out by financial intermediaries, including the credit rating institutions, for tapping the capital market reveals that these are already restricting the functioning of the para-statal agencies as also local bodies.

In several cases, the pubic bodies have been forced to pledge their regular earnings (from grants from the state, octroi etc.) as a guarantee for debt servicing. Alternately, the guarantee has come from the central or state governments to make up the loss or even ensure certain minimum rate of return to the private agency. Importantly, the projects that are likely to be financed through such arrangements happen to be commercially viable, catering to the demands of the upper or middle class, so as to ensure profitability to the investors and other stakeholders. The arrangements would, thus, lead to a situation wherein the finances generated from the common people would get diverted for financially profitable projects, or get pledged as a security for these that are likely to benefit better off sections of population. In case of an eventuality of the projects failing to generate the desired rate of profit, the public bodies would have to depend on the bail out by the government. It, thus, appears that the policy of liberating the public utilities and para-statal agencies from the regulatory and legislative controls of the state would bring the former firmly under the direct control of financial institutions by 2020. The state, however, would hardly be in a position to shake off its responsibility in case of failure of a project, despite the avowed desire of the lending institutions to treat the public bodies or local governments as corporate entities. Understandably, the funds from the capital market would be used in taking up infrastructural projects that cater to the needs of upper income groups or are located in rich colonies. This would result in a dilution of their social commitments including that of reducing regional imbalances and providing infrastructural amenities to the poor.

Despite all these, the new instruments like bonds, SDO (structured Debt Obligation) are likely to emerge as the dominant method of resource mobilisation by the year 2020. These would be more acceptable to the investors than general borrowings by the local body as it insulates the former from the general risk of financial uncertainty, faced by the issuer. Its basic strength would stem from the fact that every city would have certain areas or sectors wherein it is possible to identify commercially profitable projects. Also, they would always be able to identify a few revenue sources that are lucrative and buoyant. Escrowing the earnings from the lucrative sources can, thus, support the credit instruments floated for mobilising funds for the commercially viable projects. The instruments, issued through the state level financing institutions or special purpose vehicles (SPVs), would possibly be made further attractive for the investors by providing certain tax benefits. It is evident that the new institutional arrangements more importantly and the credit instruments would open up an opportunity for the large cities to tap new resources. These, nonetheless would seriously constrain "the local fiscal flexibility" (Mathur 1999) by curbing the autonomy of the agency to utilise its own resources for carrying out its normal functions as also meeting some exigencies, as noted above.

The requirements of the capital market for issuing bonds, furthermore, would facilitate launching of projects that have short gestation period and high financial return. Understandably, those catering to the demands of the upper and middle class consumers would attract investment funds, both from within as well as outside the country. It would be extremely difficult to get funds for social sector projects or those targetted to the poor due to their low rate of return and high risk factor. The financial intermediaries including credit rating agencies would, thus, ensure that infrastructural investment takes place in projects with assured cost recovery. They would also create pressures for bringing about an appropriate legal and administrative restructuring, at the state and city level, so that the commercial viability of such projects is guaranteed. These would seriously restrict the choice of the projects, rendering launching of schemes for say provision of water and sanitation facilities for the poor, slum improvement etc., virtually impossible in 2020. (Kundu et. al. 1999).

Until recently, ULBs had no mandate to take up infrastructure development or launch capital projects for sewerage and garbage disposal. Now, that the states are downloading these responsibilities to local bodies, these bodies would have passed these over to private companies through contractual arrangements by 2020. These bodies are very keen to subcontract solid waste disposal to private companies. Most of the ULBs are in a dire financial state and the very motive behind contracting out is to save on capital as well as current expenses. It would, however, be difficult by 2020 for them to ensure that the private companies meet the social obligations like covering the slum population, without their paying a substantial compensation. Many among the ULBs have already negotiated with the private agencies and agreed to compensate them not in cash but in kind. In most cases, the Corporations have leased out land to the companies on a long-term basis. The companies find this arrangement remunerative because they can thereby use the precious urban land for other purposes, earning often undeclared profits. We may find a number of companies taking up the job of waste disposal solely for the purpose of getting a hold on urban land for alternate uses, without going through the complex land acquisition procedures or paying their market price.

Unfortunately, it may not be possible to augment the technical competence of the city authorities to assess the cost implications of such arrangements for the people or urban economy in the long term. Given the financial situation, the state governments are unlikely to make serious efforts to create information base or provide technical assistance to ULBs for assessing the long-term implications of land deals and helping the latter in enforcing appropriate administrative safeguards. There would be no institutional network to which an ULB can look up for seeking guidance on technical matters like, the proposal to set up a garbage disposal plant by a private company. In fact, voices are likely to be raised against any such institutional "interventions". Given the enthusiasm for decentalisation, any such effort would be criticised as an attempt to undermine or impinge on the authority of the local bodies.

Despite formal pronouncements by the central and state governments from time to time and some financial support given to ULBs for these purposes, it is unlikely that the latter would succeed in reaching the poor and vulnerable in any satisfactory manner. Often the Municipal Corporations, Development Authorities and the State Level Boards have admitted to providing (directly or through private arrangements) differential levels of services to different areas or failing to cover marginal areas in the cities. Understandably, the better off localities with higher tax collections are demanding high quality facilities and even pressing for creation of separate local bodies. All these have made cross subsidisation in the provision of amenities within the cities extremely difficult in another two decades.

One is, thus, not very certain that the recent moves for liberalisation, decentralisation and invigoration of the capital market would help local bodies in undertaking infrastructural investment in the new century. These developments would, while liberating the local agencies from the control of central and state governments, place the former under some trustees or commercial banks, controlled by pure market logic. Alternately, the financial powers might also go in the hands of international "donors" and credit rating agencies who, through various innovative and complicated arrangements, can influence the expenditure pattern of municipal bodies. They would also be in a position to determine the type of projects to be undertaken and in certain cases their management system. It would be extremely important to keep a watch on the problems faced by the cities in handling their budgetary earnings and expenditures after accepting the terms and conditions for raising resources.

The adverse impact of the Constitutional Amendment on intra-urban disparity is likely to become very pronounced by 2020. Only a handful of large cities with reasonably strong economic base would have benefited from the "opportunity", opening up owing to the Constitutional Amendment. A few of these cities would possibly be able to introduce certain new taxes, increase the rates of the old ones and, at the same time, liberate themselves from the legislative and administrative controls of the state government. They might also be able to raise resources by issuing debt instruments or borrowings from international organisations, as discussed in the preceding section. It is, however, unlikely that small and medium towns would be able to benefit in a similar fashion. Consequently, the mobilisation of funds from the capital market and institutional sources would accentuate the disparity in terms of per capita expenditure and the level of amenities across the size class of urban centres during the next two decades. Within a couple of decades from now we would face serious problems due to socioeconomic segmentation in large cities. The Amendment stipulates that the ward level committees are to be constituted in all cities having more than 0.3 million people and that the former will have the powers to take decisions regarding the nature of infrastructural facilities and civic amenities, based on the capability and willingness of the residents to pay. Also, these committees will statutorily be assigned certain part of the municipal budget. The slum populations in the heart of the large cities or their peripheries, with low affordability or willingness to pay, would understandably accept a low level of civic amenities. The elite colonies, on the other hand, would be able attract private entrepreneurs and even the subsidised government programmes for improving the quality of services, based on their capacity to pay higher user charges and political connections. This would further accentuate the disparity in the availability of basic amenities across the wards and between the city and the periphery.

A large number of the industrial units would get located in the villages and small towns around the big cities in the new few decades. The reasons for their moving out of the large cities are easy availability of land, access to unorganised labour market and lesser awareness or less stringent implementation of environmental regulations.

The poor would find shelter in the "degenerated periphery", get jobs in the industries located therein or commute to the central city for work (Kundu 1989). The entrepreneurs, engineers, executives etc., associated with modern industries and business, however, would reside within the central city and travel to the periphery through rapid transport corridors. The process of segmentation, manifested in different variants in different cities, would push the incoming migrants from rural areas largely into the peripheral zones.

A number of projects would be launched with the participation of community groups and non-governmental organisations in a number of cities and towns in many of the states. This new approach while involving the community in infrastructure development may lead to withdrawal of the state or minimal budgetary support. In the absence of the states designing an institutional network for providing technical support, the poor may be left to themselves for designing and implementing the project or simply involving them in credit access and loan recovery mechanism. Despite much noise being made regarding participatory projects, these are unlikely to become an alternative to the formal slum improvement programmes launched by public agencies as there is not adequate evidence that the success cases are replicable on a large scale. Although the government is placing high weightage to provision of credit and training for the participatory projects, it may not be possible to do so through institutional sources due to bureaucratic hurdles. The major handicap in large-scale multiplication of the projects would be the slum households not possessing legal title to "their land".

Changing Structure of Urbanisation and Access to Infrastructure and Basic Amenities

The changes being introduced in the system of financing infrastructural financing have serious implications for the trend and pattern of urban growth in future years. During fifties and sixties, some kind of ceiling on the absorptive capacity of large cities was sought to be imposed through physical planning and controls on location of economic activities and urban land-use, imposed through Master Plans etc. These, however, could not restrict RU migration as there was large-scale violations in the Master Plan norms and land use restrictions. The growth urban population thus worked our as high or medium during the first four decades since Independence.

The annual exponential growth rate was very high-3.5 per cent per annum-during forties. This had prompted several scholars propound the thesis of urban explosion, dysfunctional urbanisation, urban accretion etc., basically stipulating that urban centres are growing beyond their capacity, determined by the level of infastructural facilities. The rate came down substantially during fifties and sixties but that was attributed to the adoption of a rigorous definition of urban centres in the Census of 1961. In seventies, however, the country recorded an all time high growth of 3.8 per cent. This was due not only to the existing urban centres experiencing an acceleration in their growth rates but emergence of a large number of new towns.

There has been a striking change in the urban scenario in the country since the eighties reflected in deceleration in urban growth. The annual growth rate of population recorded for urban India was as low as 3.1 per cent during 1981-91. This has gone down further to 2.7 per cent during 1991-2001. It is indeed true that the numbers of new towns identified in the last two Censuses are not as high as reported in Census 1981. The main reason for the deceleration is however, slower demographic growth recorded in existing urban centres. The declining demographic growth goes against not only the popular perception of "urban explosion" but also questions the projections made by Expert Groups set up by various government departments during eighties and nineties, besides that of the India Infrastructure Report and the Tenth Five Year Plan. Even the United Nations had projected urban growth during nineties to be close to that of the eighties. All these can now be dismissed as overestimates. The data from Population Census 2001, thus, provides an occasion to have a fresh look at the infrastructure development policy in the country and the perspective on urbanisation.

Measures of structural reform in urban sector launched informally during eighties but formalised since 1991 were expected to accelerate rural urban migration and give boost to the pace of urbanisation. The proponents of the reform often argued that linking of India with global economy would lead to massive inflow of capital from outside the country as also rise in indigenous investment. This, in turn, would give impetus to the process of urbanisation since much of the investment and consequent increase in employment would be either within or around the existing urban centres. Even when the industrial units get located in rural settlements, in a few years the latter would acquire urban status, resulting in emergence of a number of new urban centres by the year 2020.

Critics of globalisation had, however, pointed out that employment generation in the formal urban economy might not be high due to capital intensive nature of industrialisation. Also, the reduction in the rate of public sector investment in infrastructure would continue for the next couple of decades, for keeping budgetary deficits low. A cut in infrastructural investment in rural areas, coupled with open trade policy, would, in turn, slow down agricultural growth, causing high unemployment and exodus from rural areas. This would lead to rapid growth in urban population. Thus, the protagonists as also the critics of economic reform converged on the proposition that urban growth in the post liberalisation phase would be high. The data from the Population Census 2020, however, is likely to prove them both wrong as there will be significant decline in urban growth not merely due to slowing down of natural growth of population but also RU migration.

More important than the decline in the growth rate of urban population is the change in the pattern of urban growth. During 1951-

91, urban growth was generally high in relatively backward states, the states of Bihar, Uttar Pradesh, Rajasthan and Orissa and Madhya Pradesh topping the list. This is because the pace of RU migration and urbanisation were high in most of the backward states and regions that were stuck in the vicious circle of poverty. The relationship between urban growth and economic development was negative but not very strong as a few among the developed states such as, Maharashtra, Gujarat and Haryana, too, recorded high or medium growth.

The other developed states like West Bengal, Tamil Nadu and Punjab experienced low urban growth. The scenario of urban growth during this period was characterised by dualism. The developed states attracted population in urban areas due to industrialisation and infrastrctural investment. However, the backward states – particularly their backward districts and small and medium towns – also experienced rapid urban growth. This can party be attributed to government investment in the district and *taluka* headquarters, programmes of urban industrial dispersal, and transfer of funds from the states to local bodies through a need based approach. A part of RU migration in backward states could also be attributed to push factors, owing to lack of diversification in agrarian economy.

Nineties, however, make a significant departure from the earlier decades. The developed states like Tamil Nadu, Punjab, Haryana, Maharashtra and Gujarat have registered urban growth above the national average. West Bengal is the only exception whose growth rate is not particularly impressive. The backward states, on the other hand, have experienced growth either below that of the country or at the most equal to that. It may, therefore, be argued that the dynamics of urban growth would become weak and tend to get concentrated more and more in developed regions, with the exclusion of the backward states, during 2001-2021.

The decline in the rate of urbanisation and the increasing concentration pattern of growth in the next couple of decades must be examined in relation to the new systems of urban governance and methods of financing infrastructural investment. Indeed, launching of the measures of privatisation, strengthening of legal system relating to pollution and land-use etc., would adversely affect the "informal and illegal" market for land and provision of civic services. It may be noted that the functioning of these informal markets had helped the poor to find a foothold in urban centres during the first few decades since Independence. The stricter implementation of planning norms and frequent invocation of legal system for eviction etc. during the incoming decades would further slow down inmigration of rural poor. Lack of their access to basic amenities, due to a reduction in public investment on urban development and social sectors, would be yet another factor, adversely affecting the pace of urbanisation.

It is possible to predict an increase in inequality in the urban structure, along with regional imbalance in the incoming decades, as discussed above. The distribution of population in different size class of settlements, as defined by the Census, is likely to become more and more skewed. The share of Class I towns or cities, with population size of 100,000 or more, has gone up significantly from 26 per cent in 1901 to 65 per cent in 1991. The percentage share of class IV, V and VI towns, having less than 20,000 people, on the other hand, has gone down drastically from 47 to 10 only. This is largely due to the fact that the towns in lower categories have grown in size and entered the next higher category. Unfortunately, however, there has not been a corresponding increase in the number of urban centres, especially at the lower levels, through transformation of rural settlements during the period from 1901 to 1991. The slow process of graduation of large sized villages into towns, through growth of industrial and tertiary activities, would be a major problem in India's urbanisation in the next few decades. The number of additional urban centres identified by the Census of 2001 is 630 which is less than that of 1991 and much below that of 1981. The process of sectoral diversification in rural economy is so week that one would not expect more that 800 additional urban centres during the next two decades. Also, given the spatial concentration of the growth process, these new towns are likely to remain concentrated around a few large cities or regions.

The second and more important reason for the urban hierarchy to become skewed and top heavy in the year 2021 is that the larger urban centres would experience faster demographic growth as compared to smaller order settlements. The class I cities, for example, have registered an average annual growth rate of 3.0 per cent during 1981-91, which is higher than that of lower order towns.

The same pattern is noted during 1991-2001 as well. More important, the class I cities exhibit a lower disparity in their growth rates, measured through coefficient of variation, compared to those in other size classes. One would expect the former to experience relatively higher and stable demographic growth in future years as these would get more and more linked to the national and even global market. In the smaller towns that are mostly rooted in their regional economy, however, population growth would tend to be low and fluctuating over time and space. This would further reinforce the dual urban structure in the country wherein the larger cities would get integrated with higher order system and would thus share the growth dynamics at national or global economy. This would, by and large, be absent in case of smaller towns.

The higher demographic growth in large cities, compared to the smaller towns, could, at least partly, be attributed to the measures of decetralisation whereby the responsibilities of resource mobilisation and launching infrastructural projects have been given to the local bodies, as noted above. The large municipal bodies, particularly those located in developed states, tend to have a strong economic base, and this would result in their high economic and demographic growth during the next couple of decades.

It has been noted that a strong lobby is emerging, particularly in these cities, pleading for disbanding all zoning restrictions, building laws and bye-laws and making them relatively independent of state and central level controls. As a consequence, decisions regarding location of industries, change in land-use etc. would be taken expeditiously at the local level. The decentralisation of responsibilities for development planning, sought to be ushered in through the 74th Constitutional Amendment, would help this lobby. The large cities that have relatively stronger economic base would be able to benefit from this opportunity of empowerment of local governments. As a consequence, a few of these cities would be able to attract infrastructural investment and record a high population growth. The small and medium towns, on the other hand, are unlikely to benefit from this changed policy regime. All these would explain the urban structure becoming more and more top-heavy by 2020, as mentioned above.

It was pointed out earlier that planning controls, bye-laws etc. have had little impact in restricting demographic growth of large cities in the past. Interestingly, the advantages of relaxation in these controls as also systems of infrastuctural financing in the next two decades would also be taken up mostly by these cities. Due to their strong economic base and consequently high tax and non-tax revenue generating capacity, these would be in a position to put in more resources in creating basic physical conditions, necessary for attracting private investment. This would make the urban structure further top heavy as these large cities and their immediate hinterland would absorb most of the migrants.

A Perspective for Balanced Infrastructure Development and a Strategy for Intervention

It has been pointed out that launching of the programme of structural adjustment has led to a decline in public investment in infrastructure, without any compensatory gain in the contribution from the private sector. Public undertakings and para-statal agencies have become increasingly dependant on their internal resources and institutional\ private finance that have constrained their capacity to take up social sector projects. With the objective of "bringing in efficiency and accountability in their functioning", the component of subsidy has also been brought down while the cost of borrowing has gone up significantly. Furthermore, there has been increasing reliance on partnership arrangements, privatisation and community participation for launching infrastructure development projects.

It has been demonstrated in the preceding section that this changed perspective and new development strategy would accentuate disparity in the levels of infrastructural facilities across the states and size class of urban settlements by the year 2020. Understandably, the poor would get priced out of the system, organised through private or joint sector initiatives. Unfortunately, the same would be true with the delivery system under the government and semi-government undertakings as well, since these have increasingly been made to depend on institutional borrowings and capital market. All these would accentuate interurban inequality in the availability of infrastructural facilities. Furthermore, this would lead to spatial segmentation, particularly in the large cities of the country.

One may ask the question whether intervention by state in India can change the process and pattern of infrastructural development, as discussed above. The analysis of the emerging policy perspectives makes it clear that posing the issue as state versus market does not give much analytical mileage. One must ask what are the tools of intervention available to the Indian state and with what objectives these are likely to be used in the next two decades.

The changes in the system of governance and financing of infrastructural investment, brought about through structural adjustment, as recommended by several international agencies and broadly accepted by the Indian planners, envisage state's role as a facilitator. If indeed the public agencies intervene as a facilitator in the market-removing its deficiencies and saving the actors from market failures-the above scenario of growing regional imbalance would emerge, possibly with a greater ease. Given the socio-political reality in India, it is difficult for private sector to bring about the changes in the pattern of investment in infrastructure without state becoming an active partner and bringing about the required legislative and administrative changes. Indian state has indeed responded quite favourably in the past few years by ushering in the necessary changes, although the democratic structure and bureaucratic inertia have made the process slow. The government has removed impediments in the functioning of the urban land market and facilitated the necessary changes in landuse, etc. The message has come loud and clear from the Plan documents that such changes are possible and are forthcoming.

In order to make a dent on the process of infrastructural development and their impact on urban structure, it would be important to make a significant departure from the present policy of financing the investment. Based on the above overview of the trend at macro level and its negative implications, a strong case can be made for providing special assistance to the less developed states that are not in a position to allocate requisite funds for infrastructural projects.

Particularly, small and medium towns in these states need to be supported, as their economic bases are not strong to mobilise resources for the purpose. This would imply increasing the resources allocated in the public sector for development of infrastructure. There must, however, be explicit stipulations in the schemes to ensure that most of the funds go to small and medium towns.

The capacity of the state or local government to generate employment directly through anti-poverty programmes would remain limited. The past experiences suggest that there has been considerable leakage in the self-employment programmes. Banks and other financial institutions have been unwilling to give loans to the poor, as the risk of non-recovery is very high.

Also, the assets created through wage employment programmes have not contributed significantly to the development potential or long-term income generating capacity of the poor. It is, therefore, recommended that the anti-poverty programmes should primarily be directed to creation of community based infrastructure.

The state governments should take the overall responsibility of ensuring certain minimum level of the amenities to all sections of population, in different size class of urban centres, irrespective of their income levels or affordability.

For this purpose, it would, be important to set up the "standards" for the infrastructural facilities in realistic terms, based on what the country can afford. The government may, however, fulfil this responsibility by engaging\ supporting private organisations, NGOs and CBOs or strengthening the local bodies.

Constitutional amendment for decentralisation of financial powers is not adequate for augmenting resources of the local bodies for this purpose. Further, it would be erroneous to depend on the capital market or the banking sector for fulfilling the social obligations or implementing a strategy of balanced regional development. All these must be backed up by actual devolution of powers and responsibilities by state governments and their use by municipal bodies. The management capacities of these bodies need to be strengthened by giving more technical personnel and training the existing staff. Manufacturing activities at the town level are noted to exhibit a strong relationship with the availability of infrastructure and civic services. One may, therefore, argue that the provision of these services in small urban settlements, besides being a goal in itself, would help in generating non-agricultural employment and diversifying their economic base.

Much of the subsidised amenities provided through the governmental programmes during the seventies and eighties have in the past gone to a few developed regions and large cities and benefited generally the high and middle-income colonies. There is no way that this can continue in a more liberalised regime during 2001-21.

However, withdrawing government support and relegating the provision of the services to the market would have adverse consequences for regional inequality as also the micro environment\ health conditions within large cities. Institutional borrowings by public agencies (involved in the provision of the amenities) at high rates of interest, reduction in their grants from government etc. are likely to erode their capacity to invest in backward states, smaller order towns, slums and low income areas.

The relationship of urban industrial growth with that of infrastructural facilities has been positive. The government and para-statal institutions providing the facilities have unfortunately not exhibited sensitivity in favour of backward states in recent years.

The pattern is likely to become more unfavorable to the developed states that have little capacity to attract corporate capital in future years. Similarly, the large cities would succeed in cornering much of the investment that would come to infrastructure sector. The small and medium towns, on the other hand would attract private investment neither from within nor outside the country.

To counter the above trends, new programmes must be designed that can cover the entire urban hierarchy and all vulnerable sections of population. Importantly, the capacity of the people in small and medium towns in less developed states as also the urban poor to pay for basic services would remain low during the years of structural adjustment as the prospect of an increase in their real income does not seem very bright.

The programmes must, therefore, be specifically targetted and the subsidies must be made more explicit and transparent. Above all, these programmes must emerge from a long term development plan for the regional economy that gives due weightage to the infrastructural needs of all sections of population in the city as also its rural hinterland.

8

Housing and Slums

Nearly one billion people alive today – one in every six human beings – are slum dwellers, and that number is likely to double in the next thirty years, according to UN-HABITAT's new publication.

The Challenge of Slums: Global

Report on Human Settlements 2003

Unprecedented urban growth in the face of increasing poverty and social inequality, and a predicted increase in the number of people living in slums (to about 2 billion by 2030), mean that the United Nations Millennium Development goal to improve the lives of at least 100 million slum dwellers by 2020 should be considered the absolute bare minimum that the international community should aim for, according to the report to be released in October 2003. The locus of poverty is moving from the countryside to cities, in a process now recognized as the "urbanization of poverty." The absolute number of poor and undernourished in urban areas is increasing, as are the numbers of urban poor who suffer from malnutrition, say the report's authors.

This movement towards "full urbanisation", which has already been completed in Europe and in North and most of South America, means that most new population growth will be absorbed by the cities of the developing world, which will double in size by 2030. Three quarters of this growth will be in cities with populations of 1 to 5 million people, and in smaller cities of under 500 000 people.

Economic Trends at the Heart of Slum Growth

Slum formation is closely linked to economic cycles, trends in national income distribution, and in more recent years, to national economic development policies. The report finds that the cyclical nature of capitalism, increased demand for skilled versus unskilled labour, and the negative effects of globalisation– in particular, economic booms and busts that ratchet up inequality and distribute new wealth unevenly – contribute to the enormous growth of slums.

Slum development is fuelled by a combination of rapid rural-to-urban migration, spiralling urban poverty, the inability of the urban poor to access affordable land for housing and insecure land tenure. While traditional approaches to the slum problem have tended to concentrate on improvement of housing, infrastructure and physical environmental conditions, the report's authors advocate a more comprehensive approach to addressing the issue of employment for slum dwellers and the urban poor in general.

Slums are largely a physical manifestation of urban poverty, a fact that has not always been recognized by past policies aimed either at the physical eradication or the upgrading of slums. For this reason, future policies must go beyond the physical dimension of slums by addressing the problems that underlie urban poverty. Slum policies should be integrated with broader, people-focused urban poverty reduction policies that deal with the varied aspects of poverty, including employment and incomes, shelter, food, health, education and access to basic urban infrastructure and services.

Improving incomes and jobs for slum dwellers, however, requires robust national economic growth, which is itself dependent upon effective and equitable national and international economic policies, including trade. Current evidence suggests that globalisation in its present form has not always worked in favour of the urban poor and has, in fact, exacerbated their social and economic exclusion in some countries. There is abundant evidence of innovative solutions developed by the poor to improve their own living environments, leading to the gradual consolidation of informal settlements. Where appropriate upgrading policies have been put in place, slums have become increasingly socially cohesive, offering opportunities for security of tenure, local economic development and improvement of incomes among the urban poor. UN-HABITAT's Global Campaign on Secure Tenure is closely linked with policy intervention in slums. The campaign is designed to promote the commitment of Governments to providing "Adequate Shelter for All", one of the two main goals of the Habitat agenda. Providing secure tenure is seen as essential for a sustainable shelter strategy, and is a vital element in the promotion of housing rights.

In addition, the quality of urban governance plays a central role in the eradication of poverty and slums, and the creation of prosperous, more liveable cities. UN-HABITAT's Global Campaign on Urban Governance, launched in 1999, envisions and promotes "inclusive cities", cities in which everyone, regardless of their economic status, gender, race, ethnicity or religion, is enabled and empowered to fully participate in the economic and political opportunities that cites have to offer.

> *"For slum policies to be successful, the kind of apathy and lack of political will that has characterized both national and local levels of government in many countries in recent decades needs to be reversed,". "Much more political will is needed at all levels of government to confront the huge scale of slum problems that many cities face today, and will no doubt face in the foreseeable future."*

Investment in Infrastructure Key

At the core of efforts to improve the environmental habitability of slums and enhance economically productive activities is the need to invest in infrastructure – to provide water and sanitation, electricity, access roads, footpaths and waste management. Low-income housing and slum-upgrading policies need to pay attention to the financing of citywide infrastructure development. Having said that, however, the main focus of policy makers must be on poverty reduction and the up-grading of slum communities.

The report finds that upgrading existing slums is more effective than resettling slum dwellers and should become the normal practice in future slum initiatives. It goes on to state that the eradication of slums and resettlement of slum dwellers can create more problems than are solved. Eradication and relocation unnecessarily destroy a large stock of housing affordable to the urban poor and the new housing provided has frequently turned out to be un-affordable, with the result that relocated households move back into slum accommodation.

Such policies offer opportunities for more secure tenure, local economic development and improvement of livelihoods and incomes for the urban poor. They can transform the settlements in which the urban poor struggle to survive from filthy ramshackle housing developments characterised by disease and insecurity to upgraded, well-maintained homes, where families and communities can thrive.

Poverty and Slums in India – Impact of Changing Economic Landscape

Western media headlines as usual are as follows – twenty five percent of Indians live on less than a dollar a day and seventy percent live on less than two dollars a day. The forgoing was the headline of May 9, 2005 in a major international newspaper. Others headlines are not any less mischievous.

These are all meaningless analysis. It does not reflect that same amount of money has differing values in different places. A more acceptable and bit accurate description of incomes in countries is Purchase Power Parity (PPP), which is, pricing identical products and services as needed by the local population in different countries, thus establishing a new and a more equitable exchange rate. The foregoing is applicable mostly to tradable goods.

The PPP will put India's GDP at $3.7 Trillion. This will raise daily monies of twenty five percent of Indians at the lowest rung of the society to seven dollars.

The latter is still low but is much higher than the Western media would like to project. The forgoing is not the point; the point is that poverty is a major shame in India's otherwise decent, scientifically advanced, peace loving and at times turbulent image. Poverty creates slums and slums breed hopelessness and crime. Hence it needs to be tackled as an integral part of economic development.

The key question that arises-will the current hype in economical development in India alter the landscape for the very poor?

The answer is that, not much will change in next 20 to 25 years. The real impact will be felt later than twenty-five years. That is when 8% growth trajectory will take the PPP daily income of the very poor in India from seven dollars to forty dollars. By then, a $20 Trillion GDP economy (PPP basis) and $600 billion in exports (year 2001 basis) will add one hundred and fifty million jobs, of which forty to fifty million will go to the very poor segment of the society. This general prosperity will not only put food on the table but will add to better living, better housings etc. In the intervening period of 25 years, rising income levels will definitely add to the exodus from the slums to planned living areas. The forgoing also requires massive governmental effort to house people properly.

Let us examine this issue of poverty and slums in Indian cities and its relationship to the betterment of economic conditions of the masses, a bit further?

What Causes Slums in the Cities in the First Place?

It is vicious cycle of population growth, opportunities in the cities (leading to migration to the cities), poverty with low incomes, tendency to be closer to work hence occupying any land in the vicinity etc. The key reason out of all is the slow economic progress. After independence in 1947, commercial and industrial activity needed cheap labour in the cities. Plentiful was available in the rural area.

They were encouraged to come to cities and work. People, who migrated to the cities and found work, brought their cousins and rest of the families to the cities. Unable to find housing and afford it, they decided to build their shelter closer to work. First, one shelter was built, then two and then two thousand and then ten thousand and on and on. Conniving governments provided electricity and drinking water. Politicians looked at the slums as vote bank. They organized these unauthorized dwellers into a political force; hence slums took a bit of a permanent shape. More slums developed as more population moved to the cities. By mid sixties Mumbai, Kolkata, Delhi, and all other large cities were dotted with slums.

Very poor people live in slums. They are not the only one dwelling there. Fairly well to do people also reside there. They are either offspring of the slum dwellers that found education and an occupation. They have prospered but are unable to find affordable housing, hence have continued to stay in the shantytowns. Others are avoiding paying rent and property taxes. The latter is more often the case. It is not unusual that in the dirtiest of slums, where misery prevails that TV sets, refrigerators and radios are also blaring music. This is quite a contrast from the image which one gets in the media or from the opportunist politicians.

India's capital of Delhi has a million and a half out of fourteen million living in slums. Mumbai is worst with greater percentage living in slums. Other big urban centres have done no better. Newly built cities like Chandigarh and surrounding towns where shantytowns could have been avoided altogether have now slums. The forgoing is India's shame despite huge progress.

How will the Growing Economy Impact Poverty and the Slum Dwellers?

As stated above, 8% growth rate of Indian economy will push per capita GDP to $2,000 level in about twenty to twenty-five years (PPP per capita GDP will be much higher). The forgoing presupposes that the population does not explode in the near future but continue a healthy 1.5 to 2% growth. That is where the magic equilibrium of prosperity and desire to live a better life begins. These two together could end poverty and slums. With availability of affordable housing and jobs, slum dwelling is the last thought on people's mind.

On the other hand if the above does not happen then slums dwellers will triple in 25 years and so will the poverty. Delhi will have four and a half million-slum dwellers. Kolkata and Mumbai will have even bigger numbers. India's shame will have no end. To avoid that, India's economy has to remain at a high state of growth. Jobs created by the economic growth, hence higher incomes are key criteria for poverty reduction and slum elimination.

The foregoing together with the current urban renewal in progress in the urban areas today will give cities in India a new look. Higher incomes will create a demand for in-expensive housing, which will have to be met with innovative use of land and building techniques. Government provided housing would be a great failure as it has been elsewhere in the world. Instead sufficient cash has to be placed in the people's hands together with in-expensive land that people's housing program become efficient and affordable. In addition slum living has to be made unattractive with land taxes and denial of social services. Slum colonies, which opt out of current hopelessness, should get a better deal in housing which replaces the slums. This followed with rapidly growing rural economy will kill migration. That will also reduce pressure on housing.

No single policy has ever brought an end to poverty and slums. It is a concerted effort and better policies, which will end it. No country in the world has ever been able to end poverty and slums completely. That includes the richest nation of the world – USA. The point is that if economy progresses and special effort is made to uplift the poor, poverty and slums will be overtaken by better economic conditions of the people.

How did US Tackle its Slums?

US had its share of poverty and slums in around the immigrant dominated cities. New York and Boston had great amount of poverty and slums at the turn of the twentieth century. These slums worsened further with the arrival of newly liberated African-American population from Deep South. The era pictures give a glimpse of everyday life and it is not pretty. People without jobs and with no prospects crowded cities in the North. A new word, Ghetto was coined, which described these places.

Immigrant from different background or race crowded together and gave rise to Ghettos. At that time US did not have control over its economy and Civil War debt and additional monies borrowed to rehabilitate agriculture and commerce after the Civil War was unpaid. As twentieth century progressed a concerted effort was made to clean up the Ghettos and push people inland with free grant of land and promise of prosperity. Industrial Revolution, which was slow in reaching America from Europe, finally arrived. And it made the difference. It provided the much-needed jobs to the immigrants and colored. Also, free land in the West gave rise to food self-sufficiency and paying off of all Civil War and post Civil War debts. First World War gave US economy a boost and America joined the select group of countries of Europe in prosperity. Poverty by the end of the Second World War was a thing of the past. In just fifty years, i.e. by 1950, US were nation of 160 million souls, all prosperous and all well employed (forget the habitual lazy). That does not mean that all the Ghettos disappeared. They continued to exist. They exist today, but on a much lower scale. These are not eyesores.

One critical factor which eliminated slums and poverty in US was quadrupling of the US economy from 1900 to 1940. A free wheeling economy created industrial giants and a super rich class. Need for war material during the WWII resulted in creation of huge industrial infrastructure and innovation. Post war reconstruction in Europe added greater impetus to the economy. General well being of the people living in the poorer section of the cities dramatically improved. US raced ahead of Europe and are still ahead, 60 years after the WWII. In most cities, ghettos disappeared or shrunk. Urban renewal and building boom in last sixty years has completely changed the landscape of the country.

There is a parallel here. Poverty and slums in India are at the same level as they were in beginning of the twentieth century in America. Economic growth over fifty years eliminated them. It is possible in India too if the economy sustains the 8% growth trajectory.

Slums and the Great Briton

Great Briton was a great big slum before they became a colonial power in the nineteenth century. For eight hundred years prior, until 1800s, Great Briton was an agrarian society, where the lord lived happily in his Manor and Castles and the masses lived in a great squalor. Slums were everywhere. London had the biggest slums. Colonization brought prosperity and prosperity brought in a huge effort to improve the lot of the people and clean up of the cities. That is when the unemployed and slum dwellers were pushed to newly developing industrial hubs of Sheffield, Birmingham, Liverpool and Manchester. Compared to that Delhi, Kolkata were heavens. First slums in Kolkata appeared in 1850-70 as a result of systematic destruction of textile industry in Bengal and destruction of trading infrastructure in and around Kolkata. Slums elsewhere followed.

It took all the Victorian age from 1825 to 1900 to vanish poverty and slums in England. Their GDP multiplied 8 times over this period. British factories produced goods and services which were sold at profit in the in the colonies. Work for everybody in England was the cornerstone of building well-serviced cities.

The point is that reduction of poverty and slums follow closely with economic development. Faster the economic development, sooner will the poverty vanish and with it, the slums.

How did China Handle its Poverty and Slums?

Chinese had a unique way of making slums disappear from its urban centres. Permit system to live in a city or in a particular neighbourhood was introduced just after the Communist took control in 1949. That means that a migration of rural population to the urban areas in search of jobs was arrested. In addition the war ravaged eastern provinces where rural population had moved to the cities and into the slums, were emptied out. Nobody questioned Mao Tse Tung's wisdom; hence he had a free hand. People were permitted to return to their homes in the cities only after proof of their residency had been established. Outsiders were sent back to their own homes and land in the rural area. Future residency in the cities was permitted on a permit basis only.

Hence the major problem of unplanned urban squatting was prevented. Even today the foregoing policy continues. The FDI built cities of Guangdong province carry on with the permit system established in 1949. In order to move there, a person has to have a job and place to reside. The latter could be a factory provided bunk bed. This prevents urban squatters. The above is no comparison to how poverty was vanished in UK, US and elsewhere. Major economic progress in last 20 years has re-invigorated the cities with investment and reconstruction. Whether the same is true in the China's rural areas is a debatable issue. China likes to pretend that poverty has been removed. Published reports state otherwise.

Urban Renewal in India

Urban renewal is in progress in India in a big way for the last 50 years. The British starved cities in India of the funds for two hundred years. They only built regal palaces for themselves in Delhi, Shimla and Kolkota. No new funds were made available to the people to renew and rebuild, hence Moghul Delhi presented a decaying and a rundown look, when they finally left India in 1947. The problem got compounded with migration of people from rural areas. Expanding industry and commerce needed them hence migration was encouraged. Thus urban slums and squatting began in a big way. Today, some estimates place 10 to 15% of Delhi population as slum dwellers. Slums in Kolkata predate Delhi slums. So do the Mumbai slums. They all began the same way – people's livelihood was destroyed or they were invited to work in factories without adequate housing. The problem grew acute with huge population growth after 1950. From 1950 to today, cities lacked funds to renew themselves and help build additional housing. People lacked adequate jobs hence are caught in the poverty cycle.

Only recently a huge building and construction boom has started in all cities in India. Whereas governments are concentrating on building infrastructure and industrial base, private construction is building work places, shopping districts and housing for the middle class. The poor and slum dwellers are not there in any building equation. Cheap housing projects are lowest in the category. Hence slum dwelling has become a way of life.

How Long the Poor have to Wait?

If the experience elsewhere is a guide then poverty, slums and urban squat will be a diminishing phenomenon, if the rapid economic progress keeps its pace. Today we would have smaller of the slums, had economic policies of the present were in place 50 years back. Only now, all signs point to a rapidly rising GDP together with rising per capita GDP. With rise in income level, tendency to head to the slums has lessened. Die-hard slum dwellers who wish to pay no taxes and spend nothing on housing will most certainly continue to stay there. Others will prefer to move out. This is a normal phenomenon. It happened in US and elsewhere.

It will happen in India too. An economic equilibrium has not been reached in the society yet, where enough money in people's pocket will persuade them to vacate the slums. This won't we reached for another 20 to 25 years. By about middle of this period with increased availability of housing and higher incomes, the growth in slum dwelling will be arrested. Decline will begin only when much higher incomes are reached (as stated above), provided India does not make the mistake of regularizing the slums/bustees with land tenure on tenable land and other amenities. That is a sure fire method to keep the slums going. People will always wait for free grant of land ownership even if these grants never materialize. Even the possibility of this ever happening in a distant future will keep the slum dwellers in the slums.

Poverty, slums and urban squat are not going to go away in next 20 to 25 years. Reversal of this phenomenon will begin after sufficient economic progress had been made. Eight percent GDP growths is a good sign. With quadrupled GDP in 25 years, there is a good chance that the new and upcoming generation may stay away from slum dwelling. It may take another 25 years before the slums are vacated.

Just as Slums Need Cities to Survive, So do Cities Need Slums to Thrive

"Slum" is a very sticky word. It's short, simple and difficult to replace. As a result, it is, more often than not, misused in the context of cities in developing countries. A slum is a housing area that has deteriorated. It was originally applied to those parts of cities that were once respectable, even desirable, but which have gradually deteriorated as the original residents moved out to newer, better areas of town. The once "respectable" houses became rental properties, with the accommodation modified to house an increasing number of occupants. Thus a four-storey family mansion became four "flats", with each room housing one or more family.

Generally, as the living accommodation gets subdivided, the services remain the same. The kitchen and the bathroom become shared. As the property gets overloaded and overcrowded, its rental value goes down, and especially if rent controls are enforced, the rents are no longer enough to justify repairs. Thereafter, the building goes into limbo, living out a slow death. In some cases, its fortunes may revive, if the area then becomes a target for artists and architects who can spot a bargain lurking behind the makeshift partitions. If conditions are right, there may even be a general overhaul of the area as part of an urban renewal programme.

These are the slums that all cities have had to deal with. These are the slums that Engels wrote about in England over a hundred years ago, a direct outcome of the industrial revolution, and which justified government involvement in housing, and the introduction of building control regulations and planning laws. These are the slums that we would all like to get rid of.

However, there is another form of housing, especially-but not limited to-cities in developing countries. These are the "informal settlements" that house migrants to big cities. Often on the edges of the built-up areas, they provide accommodation for those that cannot afford-and are not entitled-to live in the housing provided in the planned settlements. In each country, there is a name for them: bidonville, katchi abadi, bustee, favella, barrio, kampung, that reflect either their rural character or material status.

There is no acceptable generic term for these dwellings. Informal settlements may be accurate but is not only a mouthful, but not expressive enough for many. Often, the land on which these dwellings are built does not belong to the residents. Yet it would be inaccurate to call the residents squatters as many of them have title or rights or are paying rent for the land or the structure.

Unfortunately, we tend to lump this type of housing together and refer to it as "slums". Not only are they different in origin, character and the role they play in cities, the type of response needed to deal with them is also different. It would, of course, be too much to ask of those who profess a concern for cities and their housing areas to be selective about the term they use, and we will have to make do with "slums" as the catchall. I will try to distinguish between the upwardly mobile informal settlements and the static or deteriorating slums, but insisting upon it may well be thought unnecessarily pedantic.

Are Slums Inevitable?

All cities have their slums and their informal settlements. Their extent, proportion and character vary not merely with income level but also with the socio-political or legislative environment and law-enforcement regime. Ironically, it is not so much the absence of a legal framework and its application that leads to slums and informal settlements; its very presence can frustrate efforts to pre-empt and prevent their formation and growth.

In my opinion, slums are not only inevitable, they are a mark of success of a city. The formation of slums is an integral part of the process of growth and development of a city. Only in a static (stagnant?) city does the state and status of its constituent parts remain unchanged. This is not to say that each part of a city must go through a cycle of development, deterioration and renewal, but that at any given moment of time, there are parts that have seen better days, and parts that are being newly developed or renovated. The extent of deterioration is, of course, relative, and the worst "slums" of a city may still be better than most parts of many. The decline and renovation of some cities may be piecemeal and hardly visible on the outside or to outsiders, and an area may continue to be held in esteem even as properties within it are gutted and remodelled, while retaining their facades.

However, the slums most people are against are the large-scale deterioration of structures, infrastructure and living conditions. Though not inevitable, these usually only come about in dynamic, rapidly changing cities. Thus while we should deplore the conditions in slums, we should see their formation as an indicator of urban success. They play a useful role in providing cheap (though not necessarily cheerful) housing for those who cannot or, as likely, will not, want to spend any more on housing than they possibly can. These areas in transition can facilitate migrants in the process of consolidating their transformation into citizens.

Insisting on a "city without slums", especially when no alternative housing has been developed can mean even more hardship for the very group that is so essential to urban development: the rural migrant. In most cities, the fight against slums is directed against informal settlements. Informal settlements come about because the price and the rate and scale of provision of formal settlements cannot match the demand for housing. To a large extent, this may be a failure of the formal sector, but it is also a consequence of the success of the city relative to other settlements or the rural areas.

Just as slums and slum dwellers need cities to survive, so do cities need slums to thrive. With large numbers competing for work in cities, it is easier to pay low wages. However, a worker still needs to live, and without the informal settlements, the minimal acceptable salary would really hit the pockets and the profits of the rich. In over 150 years of trying, we have yet to come up with a viable alternative. At best, the formal sector manages to house between 20 and 40 per cent of urban households. The rest manage not only without access to government handouts and subsidies, but despite the obstacles and barriers put up by government bureaucracy and law-enforcement agencies. Those who are against slums, often act on the NIMBY (not-in-my-backyard) principle. They are against the apparent chaos (read dynamism), disorderly layout (read organic), ramshackle, makeshift construction and lack of services (read rural) and the fear that these settlements are the breeding ground for crime and prostitution (read exploitation).

In fact, like other informal sector activities, informal settlements take advantage of the failures of the formal sector and use sweat equity instead of money to create a living environment, however marginal. I have yet to hear those wanting to get rid of slums and informal settlements make a plea for wages that allow affordable housing as a solution. Yet it was exactly this solution that was responsible for getting rid of the slums in cities such as Manchester and London. As long as gross wage disparities exist (making it possible for cities to employ cheap labour), slums are here to stay.

So how do we cope with the reality of slums in our cities? Following are some pointers:

- Recognize the city's need for migrants and make good use of the migrants' energy, drive, enthusiasm and willingness to make a new and better life for themselves, and thereby enriching the city at a very low cost to the city.
- Recognize the needs of migrants: their housing needs are minimal compared to their need for income and employment. This allows for the incremental development and gradual improvement of settlements without front-loading them with excessive infrastructure and construction costs.
- Support the transformation of informal settlements. All informal settlements gradually improve over time if they are not "eradicated", but some do so faster and better than others. Provide the support required to speed up the process, by providing access to financial, organizational and technical information and sources.
- Don't worry about providing land titles-the poor don't borrow from institutions that require them. "Entitle" the urban poor with the right to settle and recognize the settlement. If the settlement is in the "wrong" location, make this clear at the very beginning. Better still, indicate the "right" location and attract settlers to them.
- Don't worry about providing subsidies; not only do they end up with the better off, most poor households manage without one. Similarly, don't pay the private sector, or provide it with incentives, to provide housing or housing finance for low-income residents. Use the funds to ensure a living wage.
- Reduce building and planning controls and regulations. Instead, increase facilitation and information. The more control there is, the more likely it is that someone will be using it to exhort

money. Use the power of incentives rather than penalties to assist and guide development. Use consultation instead of confrontation and use collaboration instead of compulsion. Be flexible and creative, and be prepared to make use of serendipity.

- Learn from informal settlements about standards, layout, land development and infrastructure provision. Participate in the process of creating informal settlements and facilitate their growth and development.

As per 2001 census report the slum population of India was 42.6 million, which constitute 15 per cent of the total urban population of the country. Only 12.7 percent of total Indian towns have reported slum.

As per the data 11.2 million of the total slum population of the country is in Maharashtra followed by Andhra Pradesh 5.2, Uttar Pradesh 4.4 and West Bengal 4.1 million It is highlighted that while the slum population has increased, the number of slums recorded in the 58th round is lower than the 49th round which indicated densification of slums. The reason could be due to a number of factors, including resettlement process, the conversion of some of the non-notified slums into notified slums, measures relating to evacuation of the non-notified slums, consolidation of smaller clusters and the exhaustion of capacity within the city for the formation of new slums Since slums do not house all of the urban poor but most of the urban poor are found to be living in slums. Though there is reduction in the levels of urban poverty, the number of slum dwellers in India continues to remain high.

There is higher concentration of slum population in the large urban centres. Conversely in comparatively less developed states the phenomenon of slums is not as pronounced (Census, 2001). Since adequate, affordable formal housing supply for the urban poor is not a priority for most state Governments, squatting appears to be the only housing option for poor in near future.

land values are escalating sharply, so the Govt. needs to be play a more pro-active role to provide for the poor rather than relying solely on facilitating the 'Market driven system'. There is good potential for organising slum communities as the average size of size of slum is small. A positive correlation is found between states' proportions of notified slums and slum amenities.

Whereas, a few states have shown increase in percentage of notified slums, in several other states it has shown a decline.

Logically, rationalising the slum notification process is therefore an important step to provide access to basic services to slum dwellers. Steep increase in land values and the growing interest of reality sector in land is perhaps making less and less land available for slums. Resettlement policy should lay down guide lines to minimize development based, market induced dis-placements and insure rehabilitation of project affected persons based on human rights to adequate shelter.

State slum Legislation needs to be reviewed and revised in light of slum policies. Slum improvements are primarily State subjects. Interestingly enough, the jurisdiction of Local Body responsible for slum improvements have very limited jurisdiction over land issues. Concern for housing needs of slum dwellers can be seen throughout the five-year plans.

The Task Force set up Planning Commission (Government of India) in 1983 reviewed policies related to shelter of urban poor and. recognized Slums and squatter settlements as products of poverty and social injustice, slum improvement was emphasized by firmly linking improvement programme with security of tenure, social development programmes and house improvement loans with certain cost recovery. Until today, slums are not fully recognized and integrated in the city development framework. The National Housing & Habitat Policy of 1998 recommended slum improvement reconstruction programmes through land sharing, release of additional F.A.R. and use of Transferable development rights. The policy however stipulates that land/shelter rights provided to the poor/slum dweller would be non-transferable.

The objective of the draft National Slum Policy (2001) was to integrate slum settlements and the communities residing within them into the urban area and to strengthen the legal and policy framework to facilitate the process of slum development and improvement on a sustainable basis."

Slum improvement has broadly been done through the following channels:

- Central Government Policies for Slum Improvement and Poverty Alleviation.
- Slum Improvement in selected cities through international and bilateral aid.
- Slum Improvement & Tenure regularization through State Legislation/State.

Government Programs:

- Slum Improvement/Redevelopment projects with private sector participation.
- City specific initiatives.

Since adequate, affordable formal housing supply for the urban poor is not a priority for most state Governments, squatting appears to be the only housing option for poor in near future. At the same time, land values are escalating so sharply, that the only housing that is formally coming up is the high-end variety. Cities are getting spatially fragmented into high quality formal developments and informal areas marked by insecurity and acute deficiencies. Govt. needs to be play a more pro-active role to provide for the poor rather than relying on and facilitating the 'Market". Selection of slums for notification is done on ad-hoc basis and excludes many substandard areas resulting in gross underestimation There is a need to recognize variety of settlement typologies of slum and each informal settlement has different and unique characteristics and problems and hence needs different policies and improvement inputs.

Urban Land Market and Access of Poor

Several policy documents of the Government of India lay emphasis on land and housing for the urban poor. However, in spite of this explicit recognition in urban land use planning, right to housing remains a distant dream. The JNNURM also recognizes the need to take care of housing needs of the urban poor. Last 10 years of economic reforms have observed the following features in the land market of the world metropolitan cities:

- Increase in land and property prices in metro cities;
- Land as a resource for infrastructure projects.
- Urban land diverted or given away cheaply for higher end real estate projects and townships.
- Permitting foreign direct investment in the real estate.
- Deregulation and land use zone conversions.
- Market based solution for slums.
- Introduction of new land management tools
- Slum evictions a displacement in metro cities.

There are various land legislations in India which are related to different dimensions of land. Each of these has influenced land supply in specific contexts of city for determining access of the urban poor to land. There are ownership related legislations and use related legislations. Land legislations in general have converted urban poor or informal housing occupants as illegal occupants of land the poor access land through variety of occupancy rights for instance easement rights.

The 11th Plan approach paper lays emphasis of increasing land supply and making processes of land conversion simpler. The urban housing and habitat policy of 2007 states that shelter is one of the basic human needs next to food and clothing. The Common Minimum Programme of the UPA Government commits itself to a comprehensive programme of urban renewal and to a massive expansion of social housing in towns and cities paying particular attention to the needs of slum dwellers.

The analysis shows that there are no realistic estimates of slum population living in urban areas and there have been vast fluctuations in these estimates. For housing rights activists', urban renewal is synonymous with slum demolitions. The legislative tools of the government have been used for converting urban poor as firstly illegal residents of the city and then delegitimized them. The process of globalization has made urban land almost inaccessible to the urban poor and all national level policies are based on market solutions for the poor. Thus the hopes are dim and option fewer for the urban poor.

Basic Services for Urban Poor-Innovative Actions and Interventions

Access to land, shelter and basic services is not only essential for physical well-being but is also vital for their ability to earn a living. One of the most direct influences the city administration has on the scale and nature of poverty is in what it does with regard to provision of basic services and shelter construction and improvement. The section analyses the provisions of basic services like Water Supply, Sanitation, Solid Waste management, Road Networks connectivity, Electricity, Drainage systems etc. in slums and suggests new initiatives for improving the amenities. In order to provide water supply services of required standards in slums, it is suggested to provide individual pipe connections to each slum household, promote rain water harvesting, replace damage pipes and improve delivery pressures at public stand posts and remove illegal connections.

Similarly, to improve sanitation standards, it is suggested to construct community toilets where individual toilets are not possible, to extend sewerage networks to slum areas and connect toilet outlets with that, and community management of toilets in common places.

A demand led approach for improvement of access to public transport should be adopted. Appropriate technology should be used for developing road network depending on the geographical conditions and climate etc. and people's participation needs to be promoted in operation and maintenance of public transport.

Solar, bio-gas and non-conventional energy needs to be promoted for street lights as well as in household energy use wherever possible and feasible. Complete coverage of slum households through electric connections should be ensured.

In order to ensure proper drainage systems, flood prone habitats should be shifted to higher elevation, canal banks should be raised and protected and retaining walls constructed wherever required.

Among other initiatives, it is suggested to have people's participation in design and implementation of the basic services in slums. The responsibility of O&M should be that of the stockholders and spatial analysis and GIS based databases with decision support system should be established for all the slums in a city along with computerised land records for providing an integrated plan for basic services in the slums.

9

Urban Ecology

Urban ecology is a subfield of ecology which deals with the interaction of organisms in an urban or urbanized community and their interaction with that community. Urban ecologists study the trees, rivers, wildlife and open spaces found in cities to understand the extent of those resources and the way they are affected by pollution, over-development and other pressures· Analysis of urban settings in the context of ecosystem ecology (looking at the cycling of matter and the flow of energy through the ecosystem) may ultimately help us to design healthier, better managed communities, by understanding what threats the urban environment brings to humans. There is an emphasis on planning communities with an ecological design, by using alternative building materials and methods. This is in order to promote a healthy and biodiverse urban ecosystem.

Urban Ecosystem

This is the growth in the urban population and the supporting built infrastructure has affected both urban environments and also on areas which surround urban areas. These include semi or 'peri-urban' environments that fringe cities as well as agricultural and natural landscapes. Scientists are now developing ways to measure and understand the effects of urbananisation on human and environmental health. By considering urban areas as part of a broader ecological system, scientists can investigate how urban landscapes function and how they affect other landscapes with which they interact. In this context, urban environments are affected by their surrounding environment but also affect that environment. Knowing this may provide clues as to which alternative development options will lead to the best overall environmental outcome. CSE's urban ecosystem research is focused on:

- Understanding how cities work as ecological system
- Developing sustainable approaches to development of city fringe areas that reduce negative impact on surrounding environments
- Developing approaches to urban design that provide for health and opportunity for citizens.

Overview

By 2030 it is estimated that 60% of the global population will live in a metropolitan setting. When people become educated and engaged in ecological activities, be it the study of local birds, testing the quality of area water sources, cleaning up vacant land to create parks and gardens, or planting and caring for street trees, positive changes occur for both the people and the environment. For example, urban ecology transformations such as street tree projects increase social connections among urban residents which are the building block for public safety. The ability to enjoy, feel safe, and trust others in your community, is increased in areas where urban ecological tasks are performed, leading to better quality of life. Interactions between non-living factors, such as sunlight and water, and biological factors, such as plants and microbes, take place in all environments including cities. Concentrating humans and the resources they consume in metropolitan areas alters such things as soil drainage, water flow, and light availability. For example, sidewalks and roof tops can change an area's hydrology by increasing storm water runoff and can contribute to higher urban temperatures by storing heat energy and acting as an artificial heat sink.

There are many actions that can help reduce these problems in urban communities. Tree planting helps limit the total surface area of concrete in communities, allowing for ground-water recharge, reducing overall temperature, and helping purify air. Activities such as community gardens or home gardening in urban communities are encouraged by urban ecologists. It saves community members money, and limits demand from outside inputs into the city. Designing Green Buildings allows for less energy needed to operate commercial, industrial, or residential communities. Green buildings are designed with alternative energy sources like Solar Power or Biogas. Other examples of a green design would be increased insulation, green roofs, water collection systems, composting and recycling programs, and overall efficiency. Urban communities can support a rich and diverse ecosystem. Biodiversity is increased with the availability of natural resources to support growth. So you can support a richer biodiversity by encouraging ecological activities in communities. Depending on the location of that area, the types of organisms in that community will vary greatly. Attempting to understand the factors that make some species successful in urban environments while others perish, is a common topic of research. Urban ecology does not necessarily make value judgments about whether urban environments are 'good' or 'bad'. Rather, urban ecology allows one to see what is happening in a community and, assist in developing ways to reach the goals that one would like to see in their community.

Vertical Farming

Vertical farming is a proposed agricultural technique involving large-scale agriculture in urban high-rises or "farmscrapers". Using recycled resources and greenhouse methods such as hydroponics, these buildings would produce fruit, vegetables, edible mushrooms and algae year-round. Their proponents argue that, by allowing traditional outdoor farms to revert to a natural state and reducing the energy costs needed to transport foods to consumers, vertical farms could significantly alleviate climate change produced by excess atmospheric carbon. Critics have noted that the costs of the additional energy needed for artificial lighting and other vertical farming operations might outweigh the benefit of the building's close proximity to the areas of consumption.

Background

Dickson Despommier, a professor of environmental health sciences and microbiology at Columbia University in New York City, developed the idea of vertical farming in 1999 with graduate students in a medical ecology class. He had originally challenged his class to feed 50,000 Manhattanites using 13 acres of useable rooftop gardens. The class calculated that, by using rooftop gardening methods, only 2 percent of the 50,000 people would be fed. Unsatisfied with the results, Despommier thoughtlessly suggested growing plants indoors, vertically. The idea sparked the students' interests and gained major momentum. By 2001 the first outline of a vertical farm was introduced and today scientists, architects, and investors worldwide are working together to make the concept of vertical farming a reality. In an interview with Miller-McCune.com; Despommier described how vertical farms would function:

"Each floor will have its own watering and nutrient monitoring systems. There will be sensors for every single plant that tracks how much and what kinds of nutrients the plant has absorbed. You'll even have systems to monitor plant diseases by employing DNA chip technologies that detect the presence of plant pathogens by simply sampling the air and using snippets from various viral and bacterial infections. It's very easy to do. Moreover, a gas chromatograph will tell us when to pick the plant by analysing which flavenoids the produce contains. These flavenoids are what gives the food the flavours you're so fond of, particularly for more aromatic produce like tomatoes and peppers. These are all right-off-the-shelf technologies. The ability to construct a vertical farm exists now. We don't have to make anything new. Architectural designs have been produced by Chris Jacobs of United Future, Andrew Kranis at Columbia University and Gordon Graff at the University of Waterloo. Mass media attention began with an article by Lisa Chamberlain in *New York* magazine. Since 2007, articles have appeared in *The New York Times*· *U.S. News & World Report*· *Popular Science*· *Scientific American* and *Maxim (magazine)*, among others, as well as radio and television features.

Advantages

Several potential advantages of vertical farming have been discussed by Despommier. Many of these benefits are obtained from scaling up hydroponic or aeroponic growing methods. Others relate to vertical farming building designs that would allow the use of renewable energy sources (wind and solar) and the recycling of materials of production such as water.

Preparation for the Future

It is estimated that by the year 2050, close to 80% of the world's population will live in urban areas and the total population of the world will increase by 3 billion people. In order to feed 3 billion more people using traditional farming techniques an estimated 10^9 hectares of new farmland must be created. Scientists are concerned that this large amount of required farmland will not be available and that severe damage to the earth will be caused by the added farmland. Vertical farms, if designed properly, may eliminate the need to create additional farmland and help create a cleaner environment by using recycling techniques rather than harming the environment by using traditional farming techniques.

Increased Crop Production

Unlike traditional farming, indoor farming can produce crops year-round. All-season farming multiplies the productivity of the farmed surface by a factor of 4 to 6 depending on the crop. With some crops, such as strawberries, the factor may be as high as 30. Furthermore, as the crops would be sold in the same infrastructures in which they are grown, they will not need to be transported or refrigerated between production and sale, resulting in far less spoilages and infestations than conventional farming encounters. Research has shown that 30% of harvested crops are wasted due to spoilage and infestations·

Despommier suggests that, if dwarf versions of certain crops are used (e.g. dwarf wheat developed by NASA, which is smaller in size but richer in nutrients), year-round crops, and "stacker" plant holders are accounted for, a 30-story building with a base of a building block (5 acres) would yield a yearly crop analogous to that of 2,400 acres of traditional farming.

Protection from Weather-Related Problems

Crops grown in traditional outdoor farming suffer from the often suboptimal, and sometimes extreme, nature of geological and meteorological events such as undesirable temperatures or rainfall amounts, earthquakes, monsoons, hailstorms, tornadoes, flooding, wildfires, and severe droughts. The protection of crops from weather is increasingly important as global climate changes take place more and more rapidly. "Three recent floods (in 1993, 2007 and 2008) cost the United States billions of dollars in lost crops, with even more devastating losses in topsoil. Changes in rain patterns and temperature could diminish India's agricultural output by 30 percent by the end of the century." Because Vertical Farming provides a controlled environment, the productivity of vertical farms would be mostly independent of weather and protected from extreme weather events. Although the controlled environment of vertical farming negates most of these factors, earthquakes and tornadoes still pose threats to the proposed infrastructure, although this again depends on the location of the vertical farms.

Conservation of Resources

Each acre in a vertical farm could allow between 10 and 20 outdoor acres of farmland to return to its natural state and recover farmlands due to development from original flat farmlands.

Vertical farming would reduce the need for new farmland due to overpopulation, thus saving many natural resources currently threatened by deforestation or pollution. Deforestation, desertification, and other consequences of agricultural encroachment on natural biomes would be avoided. Because vertical farming lets crops be grown closer to consumers, it would substantially reduce the amount of fossil fuels currently used to transport and refrigerate farm produce. Producing food indoors reduces or eliminates conventional slowing, planting, and harvesting by farm machinery, also powered by fossil fuels. Burning less fossil fuel would reduce air pollution and the carbon dioxide emissions that cause climate change, as well as create healthier environments for humans and animals alike. Furthermore, vertical farms would make maximal use of the locally most efficient sources of renewable energy: farms in Iceland, Italy and New Zealand would benefit from geothermal energy, desert environments such as in the Middle East would make use of abundant solar energy (in which case the structures would have to be wider than they are tall, to maximize solar energy input), and coastal areas would benefit from wind, wave or tidal energies. Vertical Farms would also be designed to convert waste water and polluted air into clean, useable resources for crop production. Also, pure Oxygen produced by photosynthesis of the crops would be released into the environment adding to the improvement of air quality surrounding the Vertical Farm.

Organic Crops

The controlled growing environment reduces the need for pesticides, herbicides, and fertilizers. Advocates claim that producing organic crops in vertical farms is practical and the most likely production and marketing strategy.

Water Recycling

Because water recycling is more practical and economical in a controlled agricultural environment, vertical farming would use much less water than traditional farming. New York City dumps 1.4 billion gallons of "treated waste water" into its rivers daily. Vertical farming would convert this blackwater (waste) and grey water into potable water by collecting the water released into the air by evapo-transpiration. Vertical Farming supports the concept of "addressing food production in a modern city, where urban wastes, like black water will be composted, recycled and used for farming inside a standard tenement-like building. This will be expected to improve the living conditions since transportation costs in handling food supply and wastes will be greatly reduced.

The city's sewage sludge will enter a machine called "Slurry Carb", to break down the sludge into carbon and water. The remaining slurry will be burned like coal to power steam turbines that will generate electricity. Part of the sludge will be treated with chemicals to kill the bacteria and will undergo heating and drying process that will convert the treated sludge into topsoil. Water extracted will undergo bio-remediation processes using cattails, sawgrass and zebra mussels, until it becomes clean enough for agricultural use. It can also be subjected for further refinement until safe enough to be used as drinking water." Today, over 70% of the liquid fresh water on Earth is used for conventional agriculture which often pollutes the water with fertilizers and pesticides. Furthermore, by using waste-water for irrigation, vertical farming would contribute to ameliorating problems with ocean dead zones, caused by algae blooms which are in turn enhanced by runoff fertilizers

Halting Mass Extinction

Withdrawing human activity from large areas of the Earth's land surface may be necessary to slow and eventually halt the current anthropogenic mass extinction of land animals. Traditional agriculture is highly disruptive to wild animal populations that live in and around farmland and some argue it becomes unethical when there is a viable alternative. One study showed that wood mouse populations dropped from 25 per hectare to 5 per hectare after harvest, estimating 10 animals killed per hectare each year with conventional farming. In comparison, vertical farming would cause very little destruction of insects and other wildlife deaths.

Impact on Human Health

Traditional farming is a hazardous occupation with particular risks that often take their toll on the health of human labourers. Such risks include: exposure to infectious diseases such as malaria and schistosomes, exposure to toxic chemicals commonly used as pesticides and fungicides, confrontations with dangerous wildlife such as poisonous snakes, and the severe injuries that can occur when using large industrial farming equipment.

Whereas the traditional farming environment inevitably contains these risks (particularly in the farming practice known as "slash and burn"), vertical farming – because the environment is strictly controlled and predictable – eliminates them altogether. Currently, the American food system makes fast, unhealthy food cheap while fresh produce is less available and more expensive, encouraging poor eating habits. These poor eating habits lead to health problems such as obesity, heart disease, and diabetes. The increased availability of fresh produce created by a Vertical Farm would encourage healthier eating habits of the surrounding population, decreasing the occurrences of major health issues related to poor dieting.

Urban Growth

Vertical farming, used in conjunction with other technologies and socioeconomic practices, could allow cities to expand while remaining largely self sufficient. This would allow for large urban centres that could grow without destroying considerably larger areas of forest to provide food for their people. Moreover, the industry of vertical farming will provide employment to these expanding urban centres. This may help displace the unemployment created by the dismantling of traditional farms, as more farm labourers move to cities in search of work. It is unlikely that traditional farms will become obsolete, as there are many crops that are not suited for vertical farming.

Energy Production

Proponents claim that vertical farms could generate power. Methane digesters could be built on site to transform the organic waste generated at the farm into biogas which is generally composed of 65% methane along with other gasses. This biogas could then be burned to generate electricity that can either be consumed at the farm or added to the grid.

Technologies and Devices

Vertical farming relies on the use of various physical methods to become effective. Combining these technologies and devices in an integrated whole is necessary to make Vertical Farming a reality. Various methods are proposed and under research.

The most common technologies suggested are:

- Greenhouse
- Aeroponics/Hydroponics
- Composting
- Grow light
- Phytore-mediation
- Skyscraper.

Plans

Professor Despommier argues that the technology to construct vertical farms currently exists. He also states that the system can be profitable and effective, a claim evidenced by some preliminary research posted on the project's website. Developers and local governments in the following cities have expressed serious interest in establishing a vertical farm: Inchon (South Korea), Abu Dhabi (United Arab Emirates), and Dongtan (China), New York City, Portland, Ore., Los Angeles, Las Vegas, Seattle, Surrey, B.C., Toronto, Paris, Bangalore, Dubai, Abu Dhabi, Incheon, Shanghai and Beijing. The Illinois Institute of Technology is now crafting a detailed plan for Chicago. It is suggested that prototype versions of vertical farms should be created first, possibly at large universities interested in the research of vertical farms, in order to prevent failures such as the Biosphere 2 project in Oracle, Arizona.

Economic Analysis Needed

The analytical work needed to establish the feasibility of vertical farming has not been done. A detailed cost analysis including operation, transportation, fertilization and soil preparation costs, crop success rates, and health-care, recycling, renewable energy, and employment benefits is required to determine the cost effectiveness of vertical farming compared to traditional farming. The extra cost of lighting, heating, and powering the vertical farm may negate any of the cost benefits received by the decrease in transportation expenses.

Because the stacked growing surfaces of a vertical farm would receive far less sunlight than the equivalent land area in a rural farm, the vertical farm would require a significant level of artificial lighting and heating to operate in all seasons. Critics have observed that high levels of artificial lighting would be needed for crops growing in areas of the building unexposed to sunlight. Bruce Bugbee, a crop physiologist at Utah State University, believes that the huge power demands of vertical farming would be too expensive and uncompetitive with traditional farms using only free natural light. He notes that the levels of light needed by growing crops is about 100 times the amount used by people working in offices. The economic and environmental benefits of vertical farming rest partly on the concept of minimizing food miles, the distance that food travels from farm to consumer. However, a recent analysis suggests that transportation is only a minor contributor to the economic and environmental costs of supplying food to urban populations. The author of the report, University of Toronto professor Pierre Desrochers, concluded that "food miles are, at best, a marketing fad." Nevertheless, Despommier has argued that vertical farming is feasible. He estimates that, using currently available technologies, one vertical farm occupying one square city block and rising 30 stories would feed 10,000 people. In other sources, he claims this number to be up to 50,000. As of now, very little data is available to support or contradict the theory that vertical farms could be a cost effective alternative to traditional farms.

In addition, the advertised benefits of vertical farming sometimes seem to assume away other problems, even though these problems have proven difficult to solve in practice. For instance, it's not clear how to ensure that any farmland taken out of production will be turned back to nature and not developed into sprawl. Similarly, if the power needs of the vertical farm are met by fossil fuels, the environmental effect may be a net loss; even building, say, wind turbines to power the farms may not make as much sense as simply building the turbines, leaving the traditional farms in place, and burning less coal. Vertical farming may, however, make sense in countries where 100% of the electricity is generated from renewable sources, such as Iceland.

Urban Wildlife

Urban wildlife is wildlife that is able to live or thrive in urban environments. Some urban wildlife such as the house mouse are synanthropic, ecologically associated with humans. There are several different types of urban area which can support different kinds of wildlife.

Adaptation of Urban Wildlife

Urban environments can exert novel selective pressures on organisms are for example, the weed Crepis Sancta, found in France, has two types of seed, heavy and fluffy, the heavy ones land near by to the parent plant, whereas the fluffy seeds float further away on the wind. In urban environments seeds that float far will often land on infertile concrete. Within about 5-12 generations the weed has been found to evolve to produce significantly more heavy seeds than its rural relatives. Among vertebrates a case is urban great tits, which have been found to sing at a higher pitch than their rural relatives so that their songs stand out above the city noise, although this is probably a learned rather than evolved response.

African Urban Wildlife

On the Cape Peninsula near Cape Town in South Africa, human development have been encroaching on baboon habitat for years and the baboons have adapted remarkably well in raiding homes for food. Elsewhere in Africa, vervet monkeys as well as baboons will adapt to urbanisation, and similarly enter houses and gardens for food. African Penguins are also known to invade urban areas, searching for food an a safe place to breed. Simons Town, next to the popular Boulders Beach had to take action to restrict penguin movement due to the noise and damage they caused. There are reports of leopards roaming suburban areas in cities such as Nairobi, Kenya and Windhoek, Namibia.

Indian Urban Wildlife

In India, the situation is similar to Africa. Monkeys, such as langurs, will also enter cities for food, and cause havoc in food markets when they steal fruit from the vendors. In Mumbai, leopards have entered neighbourhoods surrounding the Sanjay Gandhi National Park and killed several people, as the park itself is besieged by a surrounding burgeoning population, where poaching and illegal woodcutting is rife.

British Urban Wildlife

Many towns in the UK have *Urban Wildlife Groups* which work to preserve and encourage urban wildlife. One example is Oxford.

Outside

Urban areas range from fully urban, areas have little green space and are mostly covered by paving, tarmac, or building, to suburban areas with gardens and parks. Pigeons found scavenging on scraps of food left by humans and nesting on buildings, even in the most urban areas. Rats can also be found scavenging on food. Gulls of various types will also breed and scavenge in various UK cities. A study by the bird biologist Peter Rock, Europe's leading authority on urban gulls, into the rise of herring gulls and lesser black-backed gulls in Bristol has discovered that in 20 years the city's colony has grown from about 100 pairs to more than 1,200. From a gull's point of view, buildings are simply cliff-sided islands, with no predators and lots of food nearby. The trend is the same in places as far apart as Gloucester and Aberdeen. With an endless supply of food, more city chicks survive each year; to become accustomed to urban living. They in turn breed even more birds, with less reason to undertake a winter migration.

From a study conducted on great tits living in 10 European cities and in 10 nearby forests. An analysis wasmade of the way the birds used songs to attract mates and establish territorial boundaries.

Hans Slabbekoorn, of Leiden University in Belgium, said that city birds adapt to life by singing faster, shorter, and higher pitch songs in the cities compared to forests. The forest birds sing low and they sing slow. Great tits living in noisy cities have to compete with the low-frequency sounds of heavy traffic, which means their songs go up in pitch to make themselves heard. A bird that sang like Barry White in the forest sounded more like Michael Jackson in the big city.

The advent of these animals has also drawn a predator, as Peregrine falcons have also been known to nest in urban areas, nesting on tall buildings and predating pigeons. The peregrine falcon is becoming more nocturnal in urban environments, using urban lighting to spot its prey. This has provided them with new opportunities to hunt night-flying birds and bats. Red foxes are also in many urban and suburban areas as scavengers. They will scavenge and also eat insects and small vertebrates such as pigeons and rodents. People also leave food for them in their gardens. Close proximity to avian life hasn't presented a problem for people in the past, but there are new concerns about the spread of bird flu (the H5N1 virus) via infected migratory birds. Some scientists are worried that persistent human expansion could indirectly lead to a disease pandemic of global proportions. At present, at least, your chances of getting hit by a car are far greater than your chances of contracting bird flu.

Inside Houses

Numerous animals can also live within buildings. The house mouse, a small mammal, is specialised for living alongside humans, and is found inside buildings. Insects that sometimes inhabit buildings include the cockroach, the silverfish, and many various species of small beetles.

Sewers and other Urban Underground Spaces

Sewers also contain wildlife. The most well known wildlife of sewers is rats. An escaped pet boa constrictor was found living in sewage pipes in a block of flats in Manchester. A type of worm known as the tubifex worm is quite abundant as well. There is also the London Underground mosquito.

North American Urban Wildlife

Many North American species have successfully adapted to, and are thriving in the urban environment. Famous examples include coyotes, raccoons, opossums, deer and red foxes. This has led to conflict with man, as these animals will open garbage bags in search of food, eat food left out for pets, take the pets themselves, feed on prized garden plants, dig up the lawn, and so on. Some fear coyotes may be a risk to small children and that they should not be left unsupervised in areas where coyotes are known to inhabit. While there are media accounts of alligators being found in sewer pipes and storm drains, most experts think that such 'sewer alligators' are unlikely to sustain a breeding population in such an environment.

Animals known to dwell within human habitations include house mice, cockroaches, house centipedes scutigera coleoptrata, silverfish, and firebrats

Urbanization is defined as *the process of human movement and centralization towards and into cities and urban areas*, with the associated industrialization, urban sprawl and lifestyle that brings. The world's population moved from a rural past into an urban future in 2007/2008, when, for the first time in history, more than half of the globe's population could be classified as urban dwellers. A high percentage of city-dwellers are poor, with an estimated 1 billion living in slums. Urbanization is now mainly a trend in Africa and Asia, where about 40 percent of the population is urban today. Driven by continued high population growth and slow economic growth in parts of the regions, the urban population is expected to double between 2000 and 2030, reaching 54 and 55 percent respectively. Corresponding figures for Europe and North America are 80 and 87 percent; for Latin America and the Caribbean 85 percent. All told, the world's total urban population is expected to increase from 3.3 billion in 2008 to about 6.4 billion by 2050, or about 70 percent of the world population. The world's rural population is expected to peak at 3.5 billion in 2019 and then slowly decline, to 2.8 billion in 2050. Increased urbanization will also drive the development of mega-cities with 10 million inhabitants or more. It is estimated that by 2025 there will be 27 mega-cities, 20 of these in the developing world.

New Opportunities and Challenges

Urbanization affects economic relations and social structure throughout the world. It contributes to the globalization trend, with increased cross-border trade and cross-cultural ties bringing the world closer together. At the same time, urbanization creates opportunities and challenges, not least regarding sustainability. Urbanization is not only seen as an inevitable trend. It's also considered a positive development because concentrations of people make it easier to offer basic infrastructure and public services such as education and health services. Urbanization and growth go together, and no country has ever reached middle-income status without a significant population shift from rural to urban areas. Urban environments, with close human interaction, also tend to spur innovation and economic development. In developing countries, urbanization is considered necessary to sustain growth. Urbanization represents many of the major environmental problems facing the world, however, and urban areas tend to be environmentally as well as socially unsustainable.

For the chemical and fertilizer industry, and Yara itself, there are two major aspects of urbanization that are particularly relevant: Food and health. Urbanity and food involves the connection between the urbanization process and an urban livelihood, and the production and consumption of food. In most societies, agriculture has been responsible for laying the economic foundations for development, whereas urbanization lays the foundation for the next step in economic development, industrialization. Industrial activities are mostly located in urban areas, or the establishment of industrial enterprises spurs development of urban centres. Where land and water are scarce, urban areas compete with agriculture. Farmers can suddenly find themselves outbid for land by industrial firms, jeopardizing the production of food.

Food security is a major challenge closely connected to urbanization. The impact of urbanization on agriculture is also connected to the consumption patterns of city populations. Rising incomes lead to higher consumption and increased pressure on natural resources, especially in developed countries. Urban consumption may be a more imminent problem than the actual urban concentrations, causing a substantial urban footprint. Urbanity and health involves the connection between urban life and the living conditions affecting human health. Although urbanization allows more accessibility to health services, it also creates health hazards. In poor parts of the cities, health problems include inadequate water and sanitation, limited or no waste disposal and poor air quality, as well as crowded living conditions and general poverty. In such urban areas the air, land and water are often contaminated, spreading disease. In cities in the more affluent parts of the world, health hazards resulting from urbanization are mainly connected to air pollution, as well as crime, traffic and lifestyle.

Some problems connected to the urban physical environment affect virtually everyone, particularly air pollution. The burning of fossil fuels from transportation, industry and energy production is the main culprit regarding outdoor urban air pollution. Another health hazard common in, but not exclusive to, the cities is connected to lifestyle and consumption patterns, including dietary changes and obesity.

The Changing Landscape: Ecosystem Responses to Urbanization and Pollution Across Climatic and Societal Gradients

Urbanization, an important driver of climate change and pollution, alters both biotic and abiotic ecosystem properties within, surrounding, and even at great distances from urban areas. As a result, research challenges and environmental problems must be tackled at local, regional, and global scales. Ecosystem responses to land change are complex and interacting, occurring on all spatial and temporal scales as a consequence of connectivity of resources, energy, and information among social, physical, and biological systems. We propose six hypotheses about local to continental effects of urbanization and pollution, and an operational research approach to test them.

This approach focuses on analysis of "megapolitan" areas that have emerged across North America, but also includes diverse wildland-to-urban gradients and spatially continuous coverage of land change. Concerted and coordinated monitoring of land change and accompanying ecosystem responses, coupled with simulation models, will permit robust forecasts of how land change and human settlement patterns will alter ecosystem services and resource utilization across the North American continent. This, in turn, can be applied globally.

Beyond climate, land use – and its manifestation as land-cover change and pollution loading – is the major factor altering the structure, function, and dynamics of Earth's terrestrial and aquatic ecosystems.

Urbanization, in particular, fundamentally alters both biotic and abiotic ecosystem properties within, surrounding, and even at great distances from urban areas. Around the world, rates of land change will increase greatly over the next 20–50 years, as human populations continue to grow and migrate. The nature, pattern, pace, and ecological and societal consequences of land change will vary on all spatial scales as a result of spatial variation in human preferences, economic and political pressures, and environmental sensitivities. To respond, we must determine how variables influence land change and ecosystem properties at multiple interacting scales, and understand feedbacks to human behaviour.

Human social and economic activities drive land change at all scales, and may enhance or hinder the movement of materials via wind, water, and biological and social vectors, sometimes in surprising ways that cut across scales (Kareiva *et al.* 2007; Peters *et al.* [2008] in this issue). For example, individual human decisions can influence regional dynamics within a continent when many people respond similarly to the same economic or climatic driver; the Dust Bowl in the North American prairies during the 1930s is a historical example of such cumulative effects (Peters *et al.* 2004). Individual decisions can also influence broad-scale land-change dynamics on other continents; for example, a switch to soybean production in South America is being driven by market demand from China. In turn, the changes wrought by humans produce ecosystem dynamics that feed back to influence resource availability and human well-being. Human responses may ameliorate or exacerbate these effects. Thus, there are complex interactions and feedbacks between the direct manifestations of human activity and their diverse ecological consequences, across a range of interacting spatial and temporal scales. Here, we consider land change (especially urbanization) and pollution arising from human activities. Ecosystem responses to these "press" events (i.e. continual or increasing stresses on ecosystems over relatively long time frames) occur on local, regional, and continental scales, as a consequence of connectivity among resources, energy, and information in social, physical, and biological systems. For example, urban areas are both sources and recipients of atmospheric and aquatic pollutants. Eastern landscapes of the US receive depositions of air pollutants from the industrial Midwest, and small streams across the country receive pollutant loads (e.g. nitrate, ammonium) from intensive agriculture and concentrated feedlots. Meanwhile, the entire continent receives particulate and chemical inputs, borne in the upper atmosphere from distant global sources, including China and northern

Africa. Urban areas are also foci for species introductions (Hope *et al.* 2003; Crowl *et al.* [2008] in this issue). There are many important two-way interactions between urban processes and climate that further complicate responses at multiple scales. The specter of sea-level rise and more frequent and severe hurricanes resulting from regional and global climate change is particularly important for urban ecosystems, as they tend to be located near coastlines. Locally, changes in albedo, evapo-transpiration, and surface energy balance in developed areas may exacerbate global warming through urban heat island and oasis effects (Arnfield 2003; Kalnay and Cai 2003). Dust generation from construction within urbanizing areas may be enhanced by drought. These urban dynamics may contribute to meso-scale and global climate change, through massive greenhouse-gas emissions and radiative forcing of non-greenhouse gases (Pielke *et al.* 2002), and by alteration of rainfall patterns (Cerveny and Balling 1998). Profound structural modification of streams and rivers, coupled with changes in impervious surfaces, affect hydro-ecology in, and downstream from, cities and suburbs (Paul and Meyer 2001).

The goals of this review are: (1) to demonstrate that interactions among component parts of landscapes (e.g. urban, rural, wildland), and interactions across scales from local to continental, are mediated by vectors of water, wind, organisms, and people, and (2) to provide an operational framework for conducting continental-scale research on land change and pollution in social–ecological systems. Key scientific questions that can be addressed by this framework relate to the ecological consequences of land change and human–environment dynamics, both as drivers and responders, as well as the origins and fates of pollutants at multiple, interacting scales.

Regional variation in ecosystems arises as a result of different combinations of climate, vegetation, and geomorphology. Both today and over the course of human history in North America, this variation is perceived and responded to by people who make choices about where to settle and how to use the land. There are therefore recognizable regional differences in settlement density, current types and intensities of land use, land-use legacies, and rates and patterns of urban–suburban growth. As a further consequence of these continental-scale differences, diffuse, "non-point" pollution coalesces into distinct hotspots or source regions, such as large urban agglomerations or zones of intensive agriculture.

In addition to the background template of natural systems, economic and cultural drivers influence human settlement patterns. The first wave of European settlers migrating across North America brought introduced Eurasian species and agricultural methods, initiating continent-wide ecological transformations. Today, the Southwest is a particularly important recipient region for Mexican immigrants, owing to geographical proximity and cultural and environmental similarities. Over the past century, the upper Midwest has been a magnet for northern Europeans, due to historical timing and the similarity of the region to these immigrants' native lands and climate.

Given biophysical and social influences on urbanization, ecosystem responses are likely to exhibit regional differences. Classifying urbanizing regions based on both social and bio-geophysical variables could form the basis for continental-scale comparisons of urbanization and resulting ecosystem responses. We expect that the nature and strength of feedbacks among urban–suburban land-use change and ecosystem biogeochemistry, hydro-ecology, and biodiversity will vary across the climatic, societal, and ecological settings that characterize these strongly contrasting regions.

Understanding atmospheric and aquatic transport processes and pollution generation has given atmospheric and aquatic scientists a strong working knowledge of the patterns of pollutant distribution at continental and subcontinental scales. Yet, the ability to predict how connectivity across widely separated regions will lead to ecosystem responses to these patterns hinges upon a coordinated observation network distributed across pollutant gradients, coupled with experiments to identify mechanisms. At the continental scale, two hypotheses could be tested with such a network.

Hypothesis 1: Human socio-demographic changes are the primary drivers of land-use change, urbanization, and pollution at continental and sub-continental scales; in turn, these patterns are influenced by a continental template of climate and geography. We expect major land-use changes associated with urbanization and sub-urbanization, leading to spatial redistribution and transformation of energy and material resources. These changes include both the agglomeration of major US cities into megapolitan regions and the spread of housing into rural areas and wildlands. This land-use change will be geographically uneven and disproportionately associated with the southern and western regions of the US, requiring large appropriations and re-distributions of limiting resources such as water and nutrients. However, even in areas experiencing low population growth, the spatial expansion of urban and suburban land uses is much greater than the rate of population increase, due to a continuing pattern of declining developmental density and increasing land appropriation per capita (Theobald 2005).

Hypothesis 2: Human activities, their legacies, and the environmental template interact with gradients of air pollution and nitrogen (N) loading to produce substantial variation in ecosystem patterns and processes, from sub-continental to regional scales. We expect pollution from urban and agricultural areas to influence ecosystem structure in profound ways. Emissions of nitrogen oxides, ozone, volatile organic compounds, other reactive gases, and aerosols derive from combustion sources (e.g. vehicles, power plants) in urbanized and urbanizing areas. Ammonia emissions are high in intensively fertilized agricultural and urban regions. Dust and aerosols are produced both from agricultural and urban construction activities, and as secondary products of reactive atmospheric chemistry. The impacts of these pollutants will occur both near emission sources and many hundreds to thousands of kilometers away, as a result of long-range transport and atmospheric chemistry. For example, excess ammonium and nitrate emissions from combustion and fertilization in the Midwest are implicated in chronically elevated reactive N-loading to sensitive ecosystems (such as high-elevation forests) in the Northeast and mid-Atlantic states. Nitrogen loading and ozone exposure cause changes in plant chemistry, photosynthesis, and ecosystem carbon balance in sensitive ecosystems (Aber *et al.* 1991). As transport and deposition of emissions continues, high N loading and air pollution (especially ozone exposure) may produce similar changes in less sensitive systems. Additional responses at these and larger scales may include shifts in dominant plant species (Arbaugh *et al.* 2003; Fenn *et al.* 2003; Stevens *et al.* 2004), export of nitrates and acidity to streams, rivers, and estuaries, coastal electrification and harmful algal blooms (NRC 2000), and, possibly, increased invasiveness by N-demanding species (e.g. hybrid cattails, Eurasian *Phragmites* genotypes, winter annual grasses; Ehrenfeld 2003; Fenn *et al.* 2003). The megapolitan concept provides an operational framework for predicting urbanization at the broadest scale.

The phenomenon of urbanization is not restricted to the largest cities, however; although most people live in large cities (UNEP 2006), there are many more small cities than large ones. Diverse patterns of human settlement prevail across North America, from highly urbanized islands to sparsely populated forestland at high latitudes. Many gradients expressing these differences can be identified: cities from small to large, variable housing density, differences in the size of urban footprints, a shift from older cities to more suburban landscapes – all represent contexts that will affect the way that urbanization plays out. By studying contrasts or gradients between urban and wildland areas within regions and at local scales, scientists can develop a more comprehensive understanding of the ecosystem effects of urbanization and its feedbacks to society and management.

The gradients we propose in this paper differ from the original urban–rural gradient paradigm described by McDonnell and Pickett (1990). A useful way to conceptualize the difference is to view the continental gradients as a collection of urban–rural gradients, each associated with individual metropolitan areas. Understanding the extent to which differences in human–ecosystem interactions along this collection of urban–rural gradients can be attributed to their contexts will advance urban ecology. Generation and transport of pollutants at local to regional scales are well studied because of requirements for compliance with national air-and water-quality standards However, we do not know how the effects of air and water pollution from different human activities vary according to spatial context. For example, the concept of N saturation was developed for forest ecosystems (Aber *et al.* 1998) and is less frequently evaluated in grasslands, deserts, polar and alpine ecosystems, or lakes and streams.

The connectivity framework provides a useful context for understanding, within regions or even in local areas, how land change and pollution-generating activities affect nearby ecosystems. Hypotheses at this scale reflect connectivity both within and between urban areas, and with recipient ecosystems that are linked to them via wind or water vectors. We suggest three hypotheses that draw attention to differences and similarities across gradients in urban ecosystem structure and function and connectivity with the surrounding local–regional environment.

Hypothesis 3: Within urbanizing regions, landscape alteration and management result in a relative homogenization of form and function of urban land cover across climate zones.

Regardless of setting, urban ecosystems are strongly engineered by their inhabitants and may share similarities, despite great geographical or climatic differences (e.g. Walsh *et al.* 2005). For example, similar horticultural species are introduced in contrasting urban regions across North America. Redistribution of water and nutrients in urban landscapes may reduce differences between xeric and mesic regions, relative to the dramatic differences between corresponding wildland ecosystems. New conceptual models of social–ecological processes are needed to integrate causes and effects of development patterns and management choices on urban ecosystem function.

Hypothesis 4: Urbanization will generally increase connectivity via wind and animal vectors, but will disrupt connectivity via water vectors, especially at local to regional scales.

Urbanization generates air pollutants that connect human settlements to adjoining wildland ecosystems. We therefore expect to see increased deposition of pollutants downwind and at potentially large distances from urban areas (Cooper *et al.* 2001). In addition, urban areas are a major source of the greenhouse-gas emissions underlying global changes in climate (Pataki *et al.* 2006). Wind transport of nitrogen, dust, and ozone from cities to outlying areas will alter plant productivity, ecosystem nutrient retention, and plant and microbial communities (Fenn *et al.* 1999).

People also move both plants and animals. Comparison of species invasions and extinctions among land uses will show increased connectivity associated with human settlement for some species, although cities can also affect migration patterns by fragmenting habitat. Finally, because humans drastically modify water delivery and supply systems (e.g. streams, ground-water), connectivity via water will be disrupted, with dramatic consequences for aquatic ecosystems. Some hydrologic connections will be increased as a result of urbanization (e.g. transport of water from source areas to cities, dispersal of invasive species along water corridors, sheet flow on impervious surfaces), while other hydrologic connections will be reduced (e.g. instead of long, slow flow paths from uplands to streams via ground-water, urban storm water infrastructure creates new, short, fast flow paths that decrease ecological coupling between terrestrial and aquatic components of the landscape; Grimm *et al.* 2004).

Hypothesis 5: Humans fundamentally change biogeo-chemical inputs, processing, flow paths, and exports in areas undergoing development.

Research on urban ecosystems has expanded over the past decade and has seen some synthesis, yet this body of research consists largely of a collection of case studies. A continental network of urban sites would reveal how the major biogeo-chemical cycles are being altered by human activities (e.g. Kaye *et al.* 2006). We expect new, scaled models of biogeochemistry to emerge from observations of suburban, urban fringe, and exurban terrestrial and aquatic ecosystems.

The fact that we have insufficient data on too few cities has precluded comparison of patterns of resource imports and their transformations within cities, which would allow us to test this hypothesis.

At expanding urban fringes, we expect to observe changes in hydrologic balance and nutrient export as urban residents modify matter and energy fluxes through fuel, water, and fertilizer use, and as build-out increases impervious surface cover and modifies flow paths. Mercury, volatile organic compounds, endocrine disruptors, antibiotics, and nitrogenous pollutants released into streams from wastewater treatment plants or storm drainage will change species composition, nutrient retention capacity, and productivity of aquatic ecosystems. These processes have not been comprehensively evaluated across cities.

The influence of human land use and management on connectivity and ecosystems varies with spatial scale and region. Gradients in atmospheric deposition of N and sulfur at continental scales result from prevailing air-transport patterns between source regions (e.g. industrial corridors, transportation hubs, agricultural regions) and sink regions (e.g. rural regions, wildlands, natural areas). Coastal and freshwater electrification can be traced to upland agricultural activities in the Midwest and Gulf of Mexico, and to urbanization and atmospheric deposition in the Northeast.

Urban thermal regimes vary compared to their surroundings, owing to the increased heat capacity of the infrastructure coupled with altered evapo-transpiration, which is reduced in eastern cities relative to natural ecosystems, and enhanced in irrigated, semi-arid cities. However, the cross-scale interactions of urban heat islands with regional and global climate change are unknown. Sharp regional gradients in atmospheric and aquatic pollutants originating at urban point sources are superimposed on broader continental gradients of climate and long-range atmospheric or riverine transport of materials.

Perhaps most importantly, the scales at which human decision-making and actions occur are often inconsistent with the scales at which ecosystems are changing (Cumming *et al.* 2006). This mismatch in scale may be true both for causative action (e.g. automobile use by individuals and global atmospheric forcing of increased CO2) and corrective action (e.g. amelioration of electrification by point-source wastewater treatment). We offer the following hypothesis.

Hypothesis 6: (a) Urbanizing regions will be less vulnerable than wildland ecosystems to many broad-scale, directional changes in climate due to the capacity of humans to modify their environment, and cities' access to political power and resources. However, (b) urbanizing regions will be more vulnerable than rural and wildland ecosystems to extreme events, because of the greater concentration of people and infrastructure that cannot be moved or modified over the short term. In addition, (c) efforts by urbanizing regions to adjust to change will place added stress on rural and wildland ecosystems that are connected to cities due to greater resource exploitation.

Because the vast majority of the North American population lives in urban areas, the impacts of climate change on cities are of great interest. Urban areas and their institutions are able to adjust to directional and even some relatively abrupt changes, for example by increasing water supply during droughts or by strengthening infrastructure in response to the threat of hurricanes or sealevel rise.

In addition, urbanization has a profound effect on local climatic conditions. Large urban areas essentially create their own climate: lighter winds, less humidity, more or fewer rainstorms compared to surrounding rural areas. Moreover, urban engineering, conservation, and landscaping alternatives allow urban residents to limit the variability of the climate that they experience.

10

Urban Planning

Urban, city, and town planning integrates land use planning and transport planning to improve the built, economic and social environments of communities. Regional planning deals with a still larger environment, at a less detailed level. Urban planning can include urban renewal, by adapting urban planning methods to existing cities suffering from decay and lack of investment.

History

As an organized profession, urban planning has only existed for the last 60 years. However, most settlements and cities show forethought and conscious design in their layout and functioning. Agriculture and other techniques facilitated larger populations than the very small communities of the Paleolithic. It may have caused stronger, more coercive governments at the same time. The pre-Classical and Classical ages saw a number of cities laid out according to fixed plans, though many tended to develop organically. Designed cities were characteristic of the totalitarian Mesopotamian, Harrapan, and Egyptian civilizations of the third millennium BCE.

Distinct characteristics of urban planning from remains of the cities of Harappa, Lothal and Mohenjo-daro in the Indus Valley Civilization (in modern-day northwestern India and Pakistan) lead archeologists to conclude that they are the earliest examples of deliberately planned and managed cities. The streets of these early cities were often paved and laid out at right angles in a grid pattern, with a hierarchy of streets from major boulevards to residential alleys. Archaeological evidence suggests that many Harrapan houses were laid out to protect from noise and enhance residential privacy; also, they often had their own water wells for probably both sanitary and ritual purposes. These ancient cities were unique in that they often had drainage systems, seemingly tied to a well-developed ideal of urban sanitation.

Ur, located near the Euphrates and Tigris rivers in modern day Iraq also had urban planning in later periods. The Greek Hippodamus (c. 407 BC) is widely considered the father of city planning in the West, for his design of Miletus; Alexander commissioned him to lay out his new city of Alexandria, the grandest example of idealized urban planning of the Mediterranean world, where regularity was aided in large part by its level site near a mouth of the Nile.

The ancient Romans used a consolidated scheme for city planning, developed for military defence and civil convenience. The basic plan is a central forum with city services, surrounded by a compact rectilinear grid of streets and wrapped in a wall for defence. To reduce travel times, two diagonal streets cross the square grid corner-to-corner, passing through the central square. A river usually flowed through the city, to provide water, transport, and sewage disposal. Many European towns, such as Turin, still preserve the remains of these schemes. The Romans had a very logical way of designing their cities. They laid out the streets at right angles, in the form of a square grid. All the roads were equal in width and length, except for two. These two roads formed the center of the grid and intersected in the middle. One went East/West, the other North/South. They were slightly wider than the others. All roads were made of carefully fitted stones and smaller hard packed stones. Bridges were also constructed where needed. Each square marked by four roads was called an *insula*, the Roman equivalent of modern city blocks.

Each insula was 80 yards (73 m) square, with the land within each insula divided. As the city developed, each insula would eventually be filled with buildings of various shapes and sizes and would be crisscrossed with back roads and alleys. Most insulae were given to the first settlers of a budding new Roman city, but each person had to pay to construct their own house.

The city was surrounded by a wall to protect the city from invaders and other enemies, and to mark the city limits. Areas outside of the city limits were left open as farmland. At the end of each main road, there would be a large gateway with watchtowers. A portcullis covered the opening when the city was under siege, and additional watchtowers were constructed around the rest of the city's wall. A water aqueduct was built outside of the city's walls.

The collapse of Roman civilization saw the end of their urban planning, among many other arts. Urban development in the Middle Ages, characteristically focused on a fortress, a fortified abbey, or a (sometimes abandoned) Roman nucleus, occurred "like the annular rings of a tree" whether in an extended village or the center of a larger city. Since the new center was often on high, defensible ground, the city plan took on an organic character, following the irregularities of elevation contours like the shapes that result from agricultural terracing.

The ideal of wide streets and orderly cities was not lost, however. A few medieval cities were admired for their wide thoroughfares and other orderly arrangements, but the juridical chaos of medieval cities (where the administration of streets was sometimes hereditary with various noble families), and the characteristic tenacity of medieval Europeans in legal matters, prevented frequent or large-scale urban planning until the Renaissance and the enormous strengthening of all central governments, from city-states to the kings of France, characteristic of that epoch.

Florence was an early model of the new urban planning, which rearranged itself into a star-shaped layout adapted from the new star fort, designed to resist cannon fire. This model was widely imitated, reflecting the enormous cultural power of Florence in this age; "[t]he Renaissance was hypnotized by one city type which for a century and a half— from Filarete to Scamozzi— was impressed upon utopian schemes: this is the star-shaped city". Radial streets extend outward from a defined center of military, communal or spiritual power.

Only in ideal cities did a centrally-planned structure stand at the heart, as in Raphael's *Sposalizio* of 1504 (*illustration*); as built, the unique example of a rationally-planned *quattrocento* new city center, that of Vigevano, 1493–95, resembles a closed space instead, surrounded by arcading.

Filarete's ideal city, building on hints in Leone Battista Alberti's *De re aedificatoria*, was named "Sforzinda" in compliment to his patron; its twelve-pointed shape, circumscribable by a "perfect" Pythagorean figure, the circle, takes no heed of its undulating terrain in Filarete's manuscript. And, all this occurred in the cities, but ordinarily not in the industrial suburbs characteristic of this era, which remained disorderly and characterized by crowded conditions and organic growth.

Following the 1695 bombardment of Brussels by French troops of King Louis XIV, in which a large part of the city center was destroyed, Governor Max Emanuel proposed using the reconstruction to completely change the layout and architectural style of the city. His plan was to transform the medieval city into a city of the new baroque style, especially modelled on Turin, with a logical street layout, with straight avenues offering long, uninterrupted views flanked by buildings of a uniform size. This plan was opposed by the residents and municipal authorities, who wanted a rapid reconstruction, had no resources for grandiose proposals, and resented what they considered the imposition of a new, foreign, architectural style. In the actual reconstruction, the general layout of the city was conserved, but it was not completely identical to that before the cataclysm. Despite the necessity of rapid reconstruction and the lack of financial means, authorities did take several measures to improve traffic flow, sanitation and the general aesthetics of the city. Many streets were made as wide as possible to improve traffic flow.

In the 1990s, the University of Kentucky voted the Italian town of Todi as ideal city and "most livable town in the world", the place where man and nature, history and tradition come together to create a site of excellence. In Italy, other examples of ideal cities planned according to scientific methods, are: Urbino, Pienza, Ferrara, San Giovanni Valdarno, San Lorenzo Nuovo.

Many cities in Central American civilizations also planned their cities, including sewage systems and running water. In Mexico, Tenochtitlan, was the capital of the Aztec empire, built on an island in Lake Texcoco in what is now the Federal District in central Mexico. At its height, Tenochtitlan was one of the largest cities in the world, with close to 250,000 inhabitants.

Shibam in Yemen features over 500 tower houses, each one rising 5 to 11 storeys high, with each floor being an apartment occupied by a single family. The city has some of the tallest mudbrick houses in the world, with some of them being over 100 feet high (over 30 meters). In developed countries (Western Europe, North America, Japan and Australasia), planning and architecture can be said to have gone through various stages of general consensus in the last 200 years.

Firstly, there was the industrialised city of the 19th century, where control of building was largely held by businesses and the wealthy elite. Around 1900, there began to be a movement for providing citizens, especially factory workers, with healthier environments. The concept of garden cities arose and several model towns were built, such as Letchworth and Welwyn Garden City, the world's first garden cities, in Hertfordshire, UK. However, these were principally small scale in size, typically dealing with only a few thousand residents.

It wasn't until the 1920s that modernism began to surface. Based on the ideas of Le Corbusier and utilising new skyscraper building techniques, the modernist city stood for the elimination of disorder, congestion and the small scale, replacing them instead with pre-planned and widely spaced freeways and tower blocks set within gardens.

There were plans for large scale rebuilding of cities, such as the *Plan Voisin* (based on Le Corbusier's Ville Contemporaine), which proposed clearing and rebuilding most of central Paris. No large-scale plans were implemented until after World War II however. Throughout the late 1940s and 1950s, housing shortages caused by wartime destruction led many cities to subsidize housing blocks. Planners used the opportunity to implement the modernist ideal of towers surrounded by gardens. The most prominent example of an entire modernist city is Brasilia, constructed between 1956 and 1960 in Brazil.

Reaction

By the late 1960s and early 1970s, many planners realized that modernism's clean lines and lack of human scale also sapped vitality from the community. The symptoms were high crime rates and social problems. Modernism ended in the 1970s when the construction of the cheap, uniform tower blocks ended in most countries, such as Britain and France. Since then many have been demolished and replaced by more conventional housing. Rather than attempting to eliminate all disorder, planning now concentrates on individualism and diversity in society and the economy. This is the post-modernist era.

Minimally-planned cities still exist. Houston is a large city (with a metropolitan population of 5.5 million) in a developed country, without a comprehensive zoning ordinance. Houston does, however, restrict development densities and mandate parking, even though specific land uses are not regulated. Also, private-sector developers in Houston use subdivision covenants and deed restrictions to effect land use restrictions resembling zoning laws. Houston voters have rejected comprehensive zoning ordinances three times since 1948. Even without traditional zoning, metropolitan Houston displays large-scale land use patterns resembling zoned regions comparable in age and population, such as Dallas. This suggests that non-regulation factors such as urban infrastructure and financing, may be at least as important as zoning laws.

Sustainable Development and Sustainability

Sustainable development and sustainability influence today's urban planners. Some planners say that modern lifestyles use too many natural resources, polluting or destroying ecosystems, increasing social inequality overheating urban heat islands, and causing climate changes. Many urban planners therefore advocate sustainable cities.

However, sustainable development is a recent, controversial concept. Wheeler, in his 1998 article, defines sustainable urban development to be "development that improves the long-term social and ecological health of cities and towns." He then sketches a 'sustainable' city's features. These include compact, efficient land use, less automobile use yet with better access, efficient resource use, less pollution and waste, the restoration of natural systems, good housing and living environments, a healthy social ecology, a sustainable economy, community participation and involvement and preservation of local culture and wisdom. As they always have, urban planners try to implement widely accepted social policies and programs. Sustainability must be widely supported by society before planning can realistically modify actual institutions and regions. Real implementations are often complex compromises.

Collaborative Strategic Goal Oriented Programming (CoSGOP) is a collaborative and communicative way of strategic programming, decision-making implementation and monitoring oriented towards defined and specific goals. Furthermore it is to be based on sound analysis of available information, shall put emphasis on stake-holder participation, is expected to create awareness among actors, and shall be oriented towards the management of development processes. It was adopted like a theoretical model as a starting point for an analysis of redevelopment processes in large urban distressed areas in European Cities.

Background of CoSGOP'

CoSGOP has been derived from goal oriented planning (Gesellschaft fur Technische Zusammenarbeit-GTZ 1988). Goal oriented planning was originally oriented towards the elaboration and implementation of projects based on a logical framework approach which was useful for embedding specific project in a wider development frame and defining its major elements.

This approach showed its weakness because its logical rules were strictly applied and the defined expert language did not encourage the actor's participation. Considering this background, CoSGOP introduced a new approach characterised by: communication with the active involvement of the stockholders and those who are to be affected by the programm; strategic planning based on the identification of strengths and weakness, opportunities and threats, as well as on scenario building and visioning; the definition of goals as the basis for action regarding the improvement process; long-term flexible programming of interventions by the different stockholders.

Element of CoSGOP

CoSGOP is not a planning method but a process model. It provides a framework for communication and joint decision-making in a structured process characterised by feed-back loops and it facilitates a learning process of all the stockholders involved. The essential elements of CoSGOP are: Analysis of stockholders (This is oriented towards identifying stockholders' perception of problems and their interest and expectations); Analysis of problems and potentials (This analysis does not only include an overview over objective problems but also of problems and potentials as perceived by stockholders); Development of goals, improvement priorities and alternatives (The definition of goals, objectives for development requires intensive communication and an active participation of the concerned stockholders); Specification of an improvement programme and main activities (This programme is based on clear priorities defined with the stockholders); Assessment of possible impacts of the improvement programme; Definition and detailed specification of key project and their implementation; Continuous monitoring of improvement activities, feed-back and adjustment of the programme (Monitoring and feed-back are key elements of learning process and for monitoring success and failure is relevant not only the technical and economic information but also the perception of the stockholders).

Application

CoSGOP has been applied in European cross-border policy programming, as well in local and regional development programming. Recently (2004), CoSGOP model has been applied in the LUDA Project and it was improved starting from analysis of European experience about urban regeneration projects.

Collaborative Planning in the United States

Collaborative Planning in the US arose in response to the inadequacy of traditional public participation techniques to provide real opportunity for the public to make the decisions affecting their communities. Collaborative planning is a method designed to empower stockholders by elevating them to the level of decision-makers through a process of direct engagement and dialogue between stockholders and public agencies designed to solicit ideas, active involvement and participation in the community planning process. Active public involvement helps Planners create better outcomes by informing them of the public's needs and preferences and by using the public's local knowledge to inform projects. When properly administered collaboration can result in more meaningful participation and better, more creative outcomes to persistent problems than traditional participation methods can achieve. It enables planners to make decisions that reflect community needs and values; it fosters faith in the wisdom and utility of the resulting project, and the community is given a personal stake in its success.

Experiences in Portland and Seattle have demonstrated that successful collaborative planning is dependent upon a number of interrelated factors: the process must be truly inclusive with all stockholders and affected groups invited to the table; the community must have final decision-making authority; full government commitment-of both financial and intellectual resources-must be manifest; participants should be given clear objectives by the Planning staff who facilitate the process by providing guidance, consultancy, expert opinions and research; and facilitators should be trained in conflict resolution and community organization.

Aspects of Planning

Urban Aesthetics

In developed countries, there has been a backlash against excessive human-made clutter in the visual environment, such as signposts, signs, and hoardings. Other issues that generate strong debate among urban designers are tensions between peripheral growth, housing density and new settlements. There are also debates about the mixing tenures and land uses, versus distinguishing geographic zones where different uses dominate. Regardless, all successful urban planning considers urban character, local identity, respects heritage, pedestrians, traffic, utilities and natural hazards.

Planners can help manage the growth of cities, applying tools like zoning and growth management to manage the uses of land. Historically, many of the cities now thought the most beautiful are the result of dense, long lasting systems of prohibitions and guidance about building sizes, uses and features· These allowed substantial freedoms, yet enforce styles, safety, and often materials in practical ways. Many conventional planning techniques are being repackaged using the contemporary term smart growth. There are some cities that have been planned from conception, and while the results often don't turn out quite as planned, evidence of the initial plan often remains.

Safety

Historically within the Middle East, Europe and the rest of the Old World, settlements were located on higher ground (for defence) and close to fresh water sources. Cities have often grown onto coastal and flood plains at risk of floods and storm surges. Urban planners must consider these threats. If the dangers can be localised then the affected regions can be made into parkland or green belt, often with the added benefit of open space provision. Extreme weather, flood, or other emergencies can often be greatly mitigated with secure emergency evacuation routes and emergency operations centres. These are relatively inexpensive and unintrusive, and many consider them a reasonable precaution for any urban space. Many cities will also have planned, built safety features, such as levees, retaining walls, and shelters.

In recent years, practitioners have also been expected to maximize the accessibility of an area to people with different abilities, practicing the notion of "inclusive design," to anticipate criminal behaviour and consequently to "design-out crime" and to consider "traffic calming" or "pedestrianisation" as ways of making urban life more pleasant.

Some city planners try to control criminality with structures designed from theories such as socio-architecture or environmental determinism. Refer to Foucault and the Encyclopedia of the Prison System for more details. These theories say that an urban environment can influence individuals' obedience to social rules and level of power. The theories often say that psychological pressure develops in more densely developed, unadorned areas. This stress causes some crimes and some use of illegal drugs. The antidote is usually more individual space and better, more beautiful design in place of functionalism.

Oscar Newman's defensible space theory cites the modernist housing projects of the 1960s as an example of environmental determinism, where large blocks of flats are surrounded by shared and disassociated public areas, which are hard for residents to identify with. As those on lower incomes cannot hire others to maintain public space such as security guards or grounds keepers, and because no individual feels personally responsible, there was a general deterioration of public space leading to a sense of alienation and social disorder.

Jane Jacobs is another notable environmental determinist and is associated with the "eyes on the street" concept. By improving 'natural surveillance' of shared land and facilities of nearby residents by literally increasing the number of people who can see it, and increasing the familiarity of residents, as a collective, residents can more easily detect undesirable or criminal behaviour. However, this is not a new concept. This was prevalent throughout the middle eastern world during the time of Mohamad. It was not only reflected in the general structure of the outside of the home but also the inside. (refer to various religious texts and archaeological sites)

The "broken-windows" theory argues that small indicators of neglect, such as broken windows and unkempt lawns, promote a feeling that an area is in a state of decay. Anticipating decay, people likewise fail to maintain their own properties. The theory suggests that abandonment causes crime, rather than crime causing abandonment.

Some planning methods might help an elite group to control ordinary citizens. Haussmann's renovation of Paris created a system of wide boulevards which prevented the construction of barricades in the streets and eased the movement of military troops. In Rome, the Fascists in the 1930s created *ex novo* many new suburbs in order to concentrate criminals and poorer classes away from the elegant town.

Other social theories point out that in Britain and most countries since the 18th century, the transformation of societies from rural agriculture to industry caused a difficult adaptation to urban living. These theories emphasize that many planning policies ignore personal tensions, forcing individuals to live in a condition of perpetual extraneity to their cities. Many people therefore lack the comfort of feeling "at home" when at home. Often these theorists seek a reconsideration of commonly used "standards" that rationalize the outcomes of a free (relatively unregulated) market.

Slums

The rapid urbanization of the last century caused more slums in the major cities of the world, particularly in developing countries. Planning resources and strategies are needed to address the problems of slum development. Many planners are calling for slum improvement, particularly the Commonwealth Association of Planners. When urban planners work on slums, they must cope with racial and cultural differences to ensure that racial steering does not occur.

Slum were often "fixed" by clearance. However, more creative solutions are beginning to emerge such as Nairobi's "Camp of Fire" program, where established slum-dwellers promise to build proper houses, schools, and community centres without government money, in return for land on which they have been illegally squatting on for 30 years. The "Camp of Fire" program is one of many similar projects initiated by Slum Dwellers International, which has programs in Africa, Asia, and South America.

A slum, as defined by the United Nations agency UN-HABITAT, is a run-down area of a city characterized by substandard housing and squalor and lacking in tenure security. According to the United Nations, the proportion of urban dwellers living in slums decreased from 47 percent to 37 percent in the developing world between 1990 and 2005. However, due to rising population, the number of slum dwellers is rising. One billion people worldwide live in slums and the figure will likely grow to 2 billion by 2030.

The term has traditionally referred to housing areas that were once relatively affluent but which deteriorated as the original dwellers moved on to newer and better parts of the city, but has come to include the vast informal settlements found in cities in the developing world. Many shack dwellers vigorously oppose the description of their communities as 'slums' arguing that this results in them being pathologised and then, often, subject to threats of evictions. Many academics have vigorously criticized UN-Habitat and the World Bank arguing that their 'Cities Without Slums' Campaign has led directly to a massive increase in forced evictions.

Although their characteristics vary between geographic regions, they are usually inhabited by the very poor or socially disadvantaged. Slum buildings vary from simple shacks to permanent and well-maintained structures. Most slums lack clean water, electricity, sanitation and other basic services.

Etymology

"Slum" was originally used mainly in the phrase "back slum," meaning a back room and later "back alley". The origin of this word is thought to come from the Irish phrase '*S lom e* meaning 'exposed vulnerable place' The Oxford English Dictionary says it may be a "cant" word of Roma (Gypsy) origin. The etymologist Eric Partridge says flatly that it is "of unknown origin."

Other terms that are often used interchangeably with "slum" include shanty town, favela, skid row, barrio, and ghetto although each of these may have a somewhat different meaning. Slums are distinguished from shanty towns and favelas in that the latter initially are low-class settlements, whereas slums are generally constructed early on as relatively affluent or possibly a prestigious communities. The term "shanty town" also suggests that the dwellings are improvised shacks, made from scrap materials, and usually without proper sanitation, electricity, or telephone services. Skid row refers to an urban area with a high homeless population and a term is most commonly used in the United States. Barrio may refer to an upper-class area in some Spanish-speaking countries, and is only used to describe a low-class community in the United States. Ghetto refers to a neighbourhood based on shared ethnicity. By contrast, identification of an area as a slum is based solely on socio-economic criteria, not on racial, ethnic, or religious criteria.

Characteristics

The characteristics associated with slums vary from place to place. Slums are usually characterized by urban decay, high rates of poverty, and unemployment. They are commonly seen as "breeding grounds" for social problems such as crime, drug addiction, alcoholism, high rates of mental illness, and suicide. In many poor countries they exhibit high rates of disease due to unsanitary conditions, malnutrition, and lack of basic health care.

A UN Expert Group has created an operational definition of a slum as an area that combines to various extents the following characteristics: inadequate access to safe water; inadequate access to sanitation and other infrastructure; poor structural quality of housing; overcrowding; and insecure residential status. A more complete definition of these can be found in the 2003 UN report titled "Slums of the World: The face of urban poverty in the new millennium?". The report also lists various attributes and names that are given by individual countries which are somewhat different than these UN characteristics of a slum.

Low socioeconomic status of its residents is another common characteristic given for a slum. In many slums, especially in poor countries, many live in very narrow alleys that do not allow vehicles (like ambulances and fire trucks) to pass. The lack of services such as routine garbage collection allows rubbish to accumulate in huge quantities. The lack of infrastructure is caused by the informal nature of settlement and no planning for the poor by government officials. Additionally, informal settlements often face the brunt of natural and man-made disasters, such as landslides, as well as earthquakes and tropical storms. Fires are often a serious problem.

Many slum dwellers employ themselves in the informal economy. This can include street vending, drug dealing, domestic work, and prostitution. In some slums people even recycle trash of different kinds (from household garbage to electronics) for a living-selling either the odd usable goods or stripping broken goods for parts or raw materials.

Slums are often associated with Victorian Britain, particularly in industrial, northern English towns, lowland Scottish towns and Dublin City in Ireland. These were generally still inhabited until the 1940s, when the government started slum clearance and built new council houses. There are still many examples left of former slum housing in the UK, however they have generally been restored into more modern housing.

Growth and Countermeasures

Recent years have seen a dramatic growth in the number of slums as urban populations have increased in the Third World. In April 2005, the director of UN-HABITAT stated that the global community was falling short of the Millennium Development Goals which targeted significant improvements for slum dwellers and an additional 50 million people have been added to the slums of the world in the past two years. According to a 2006 UN-HABITAT report, 327 million people live in slums in Commonwealth countries-almost one in six Commonwealth citizens. In a quarter of Commonwealth countries (11 African, 2 Asian and 1 Pacific), more than two out of three urban dwellers live in slums and many of these countries are urbanising rapidly.

The number of people living in slums in India has more than doubled in the past two decades and now exceeds the entire population of Britain, the Indian Government has announced.

Many governments around the world have attempted to solve the problems of slums by clearing away old decrepit housing and replacing it with modern housing with much better sanitation. The displacement of slums is aided by the fact that many are squatter settlements whose property rights are not recognized by the state. This process is especially common in the Third World. Slum clearance often takes the form of eminent domain and urban renewal projects, and often the former residents are not welcome in the renewed housing. For example, in the Philippine slums of Smokey Mountain, located in Tondo, Manila, projects have been enforced by the Government and non-government organizations to allow urban resettlement sites for the slum dwellers.

According to a UN-Habitat report, over 2 million people in the Philippines live in slums and in the city of Manila alone, 50% of the over 11 million inhabitants live in slum areas.

Moreover new projects are often on the semi-rural peripheries of cities far from opportunities for generating livelihoods as well as schools, clinics etc. At times this has resulted in large movements of inner city slum dwellers militantly opposing relocation to formal housing on the outskirts of cities. In some countries, leaders have addressed this situation by rescuing rural property rights to support traditional sustainable agriculture, however this solution has met with open hostility from capitalists and corporations. It also tends to be relatively unpopular with the slum communities themselves, as it involves moving out of the city back into the countryside, a reverse of the rural-urban migration that originally brought many of them into the city. Critics argue that slum clearances tend to ignore the social problems that cause slums and simply redistribute poverty to less valuable real estate. Where communities have been moved out of slum areas to newer housing, social cohesion may be lost. If the original community is moved back into newer housing after it has been built in the same location, residents of the new housing face the same problems of poverty and powerlessness. There is a growing movement to demand a global ban of 'slum clearance programmes' and other forms of mass evictions.

Positive Aspects of Slums

In recent year, some environmentalists such as Stewart Brand and organizations such as the United Nations Population Fund suggested that despite the poor living conditions, slums are positive both environmentally and socially. Because the slum is characterized by a very high density of housing, its environmental impact is smaller than that of diffuse rural communities. The fertility rate of new slum dwellers is below the replacement rate; this mitigates dangers associated with overpopulation that results from manpower-intensive subsistence agriculture, and frees up arable land for the nature, or more efficient industrialized agriculture. Slum dwellers also appear to have vastly better opportunities for getting jobs, starting small businesses and climbing out of poverty than rural inhabitants.

Urban Decay

Urban decay is the process whereby a previously functioning city, or part of a city, falls into disrepair and decrepitude. It may feature de-industrialization, depopulation or changing population, economic restructuring, abandoned buildings, high local unemployment, fragmented families, political disenfranchisement, crime, and a desolate, inhospitable city landscape.

Since the 1970s and 1980s, urban decay has been associated with Western cities, especially in North America and parts of Europe. Since then, major structural changes in global economies, transportation, and government policy created the economic and then the social conditions resulting in urban decay. The effects counter the development of most of Europe and North America; in countries beyond, urban decay is manifest in the peripheral slums at the outskirts of a metropolis, while the city center and the inner city retain high real estate values and sustain a steadily increasing populace.

In contrast, North American cities often experience population flights to the suburbs and exurb commuter towns, i.e., white flight. Another characteristic of urban decay is blight—the visual, psychological, and physical effects of living among empty lots, buildings and condemned houses. Such desolate properties are socially dangerous to the community because they attract criminals and street gangs, contributing to the volume of crime.

Urban decay has no single cause; it results from combinations of inter-related socio-economic conditions—including the city's urban planning decisions, the poverty of the local populace, the construction of freeway roads and rail road lines that bypass the area, depopulation by sub-urbanization of peripheral lands, real estate neighbourhood redlining, xenophobic immigration restrictions, and racial discrimination. In cities such as New York and Boston, gentrification has eased urban decay in some areas of the cities, although most U.S. cities have highly blighted areas.

Background

During the Industrial Revolution, from the late eighteenth century to the early nineteenth century, rural people moved from the country to the cities for employment in the industrial manufacturing sector of the economy, thus causing the contemporary urban population boom. However, subsequent economic change left many cities economically vulnerable. Studies such as the Urban Task Force (DETR 1999), the Urban White Paper (DETR 2000), and a study of Scottish cities (2003) posit that areas suffering industrial decline—high unemployment, poverty, and a decaying physical environment (sometimes including contaminated land and obsolete infrastructure)—prove "highly resistant to improvement".

Changes in means of transport, from the public to the private – specifically, the private motor car – eliminated some of the cities' public transport service advantages, e.g., fixed-route buses and trains. In particular, at the end of World War II, many political decisions favoured suburban development and encouraged sub-urbanization, by drawing city taxes from the cities to build new infrastructure for remote, racially-restricted suburban towns. That was the context of racial discrimination exercised as "white flight", the middle-and upper-class abandonment of U.S. cities, and the start of urban sprawl; only the non-white and the poor inhabited the cities.

After World War II, Western economies lifted tariffs and out sourced most of their manufacturing industries and businesses overseas, where foreign labour is cheaper than domestic. During the change from a manufacturing to a services economy, buying an automobile became economically feasible for most people. In the U.S., the federal government legislated discriminatory lending practices for the Federal Housing Administration (FHA) via redlining.

Later, under president Dwight D. Eisenhower, urban centres were drained further through the building of the Interstate Highway System. In North America this shift manifested itself in strip malls, suburban retail and employment centres, and very low-density housing estates.

Large areas of many northern cities in the United States experienced population decreases and a degradation of urban areas. Inner-city property values declined and economically disadvantaged populations moved in. In the U.S., the new inner-city poor were often African-Americans that migrated from the South in the 1920s and 1930s. As they moved into traditional white European-American neighborhoods, ethnic frictions served to accelerate flight to the suburbs. In Western Europe the experience differs, in that the effect was often unknowingly assisted by public sector policies designed to clear 18th-and 19th-century slum areas and movements of people out into state-subsidized, lower-density suburban housing.

On continental Europe and Oceania, the historical core of major cities has usually remained relatively affluent; it is generally the inner-city districts and the edge-of-town suburbs made up of single-class state-subsidised housing, such as the French "cites" and British council estates, which suffer the worst decay and blight. Due to higher population densities in Europe, economics dictates that extremely low-density housing would be impractical.

Examples of Decay

The car manufacturing sector was the base for Detroit's prosperity, and employed the majority of its residents. When the industry began relocating outside of the city, it experienced massive population loss with associated urban decay, particularly after the 1967 riots. According to the U.S. Census, in 1950 the city's population was around 1.85 million; by 2003, this had declined to 911,000, a loss of nearly 940,000 people (52%). In addition, the homeless population has grown, and there are many abandoned structures in Detroit.

Britain experienced severe urban decay in the 1970s and 1980s. Major cities like Glasgow, the towns of the South Wales valleys, and some of the major industrial cities like Birmingham, Manchester, Liverpool, Newcastle, and East London, all experienced population decreases, with large areas of 19th-century housing experiencing market price collapse. Large French cities are often surrounded by decayed areas. While city centres tend to be occupied mainly by middle- and upper-class residents, cities are often surrounded by large mid- to high-rise housing projects. The concentration of poverty and crime radiating from the developments often causes the entire suburb to fall into a state of urban decay, as more affluent citizens seek housing in the city or further out in semi-rural areas. In November 2005, the decaying northern suburbs of Paris were the scene of severe riots sparked in part by the substandard living conditions in public housing projects.

Remedy

The main responses to urban decay have been through positive public intervention and policy, through a plethora of initiatives, funding streams, and agencies, using the principles of New Urbanism (or through Urban Renaissance, its UK/European equivalent). Gentrification has also had a significant effect, and remains the primary means of a "natural" remedy.

In the United States, early government policies included "urban renewal" and building of large scale housing projects for the poor. Urban renewal demolished entire neighborhoods in many inner cities; in many ways, it was a cause of urban decay rather than a remedy. Housing projects became crime-infested mistakes. These government efforts are now thought by many to have been misguided. For multiple reasons, some cities have rebounded from these policy mistakes. Meanwhile, some of the inner suburbs built in the 1950s and 60s are beginning the process of decay, as those who are living in the inner city are pushed out due to gentrification.

In Western Europe, where land is much less in supply and urban areas are generally recognised as the drivers of the new information and service economies, urban regeneration has become an industry in itself, with hundreds of agencies and charities set up to tackle the issue. European cities have the benefit of historical organic development patterns already concurrent to the New Urbanist model, and although derelict, most cities have attractive historical quarters and buildings ripe for redevelopment.

In the suburban estates and cites, the solution is often more drastic, with 1960s and 70s state housing projects being totally demolished and rebuilt in a more traditional European urban style, with a mix of housing types, sizes, prices, and tenures, as well as a mix of other uses such as retail or commercial. One of the best examples of this is in Hulme, Manchester, which was cleared of 19th-century housing in the 1950s to make way for a large estate of high-rise flats. During the 1990s, it was cleared again to make way for new development built along new urbanist lines.

Reconstruction and Renewal

Urban renewal (similar to urban regeneration in British English) is a program of land redevelopment in areas of moderate to high density urban land use. Its modern incarnation began in the late 19th century in developed nations and experienced an intense phase in the late 1940s – under the rubric of reconstruction. The process has had a major impact on many urban landscapes, and has played an important role in the history and demographics of cities around the world. Urban renewal can be extremely controversial, and has often involved the destruction of businesses, the demolition of priceless historic structures, the relocation of people, and the use of eminent domain (known as compulsory purchase in the UK) as a legal instrument to take private property for city-initiated development projects.

Renewal has often added to urban sprawl and vast areas of cities have been demolished and replaced by freeways and expressways, housing projects, and vacant lots as the outcome of incomplete projects.

Urban renewal's effect on actual revitalization is a subject of intense debate. It has been seen by proponents as an economic engine and a reform mechanism, and by opponents as a regressive mechanism for enriching the wealthy at the expense of taxpayers and the poor. It carries a high cost to existing communities, and in many cases resulted in the destruction of vibrant neighborhoods.

Urban renewal in its original form has been called a failure by many urban planners and civic leaders, and has since been reformulated with a focus on redevelopment of existing communities. However, many cities link the revitalization of the central business district and gentrification of residential neighborhoods to earlier urban renewal programs. Over time, urban renewal evolved into a policy based less on destruction and more on renovation and investment, and today is an integral part of many local governments, often combined with small and big business incentives. But even in this adapted form, Urban Renewal projects are widely accused of abuse and corruption.

History

Urban renewal can be traced conceptually back to the earliest days of urban development, and often stems from a paternalistic style of governance, albeit one which often uses utilitarian rhetoric. Its potential value as a process was noted by those who witnessed the inhumane and overcrowded conditions of 19th century London, New York, Paris and other major cities of the developed world affected by the industrial revolution. From this a slum reform agenda grew by which advocates and reformers, using a doctrine of environmental determinism, argued that reforming a degraded built environment would reform its residents. Such reform could be argued on religious, national security, compassionate, economic and many other grounds. Another style of reform – for reasons of aesthetics and efficiency – could be said to have begun in 1853, with the recruitment of Baron Haussmann by Louis Napoleon for the redevelopment of Paris.

Both strands of slum abolition valued the destruction of degraded housing and other structures above the welfare of slum-dwellers who, then as now, are often dispersed and might well discover themselves to be less well-off than before a slum clearance program

History of Urban Renewal in North America

Projects such as the design and construction of Central Park, New York and the 1909 Plan for Chicago by Daniel Burnham might be considered early urban renewal projects. Similar, the efforts of Jacob Riis in advocating for the demolition of degraded areas of New York in the late 19th century might also be seen as formative urban renewal programs.

Robert Moses

The redevelopment of large sections of New York City and New York State by Robert Moses between the 1930s and the 1970s was a notable and prominent example of urban redevelopment. Moses directed the construction of new bridges, highways, housing projects, and public parks. Moses was a controversial figure, both for his single-minded zeal and for its impact on New York City.

Other cities across the USA began to create redevelopment programs in the late 1930s and 1940s. These early projects were generally focused on slum clearance and were implemented by local public housing authorities, which were responsible both for clearing slums and for building new affordable housing.

Postwar Problems and Suburban Growth

In 1944, the GI Bill (officially the Serviceman's Readjustment Act) guaranteed Veterans Administration (VA) mortgages to veterans under favourable terms, which fuelled sub-urbanization after the end of World War II, as places like Levittown, New York, Warren, Michigan and the San Fernando Valley of Los Angeles were transformed from farmland into cities occupied by tens of thousands of families in a few short years.

Title One of the Housing Act of 1949 kick-started the "urban renewal" program that would reshape American cities. The Act provided federal funding to cities to cover the cost of acquiring areas of cities perceived to be "slums." (The Federal government paid 2/3 of the cost of acquiring the site, called the "write down," while local governments paid the remaining 1/3.) Those sites were then given to private developers to construct new housing. The phrase used at the time was "urban redevelopment." "Urban renewal" was a phrase popularized with the passage of the 1954 Housing Act, which made these projects more enticing to developers, by among other things, providing FHA-backed mortgages.

Urban Destruction

Under the powerful influence of multimillionaire R.K. Mellon, Pittsburgh became the first major city to undertake a modern urban-renewal program in May 1950. Pittsburgh was infamous around the world as one of the dirtiest and most economically depressed cities, and seemed ripe for urban renewal. A large section of downtown at the heart of the city was demolished, converted to parks, office buildings, and a sports arena and renamed the Golden Triangle in what was universally recognized as a major success. Other neighborhoods were also subjected to urban renewal, but with mixed results. Some areas did improve, while other areas, such as East Liberty and Lower Hill declined following ambitious projects that shifted traffic patterns, blocked streets to vehicular traffic, isolated or divided neighborhoods with highways, and removed large numbers of ethnic and minority residents. Because of the ways in which it targeted the most disadvantaged sector of the American population, novelist James Baldwin famously dubbed Urban Renewal "Negro Removal" in the 1960s. The term "urban renewal" was not introduced in the USA until the Housing Act was again amended in 1954. That was also the year in which the U.S. Supreme Court upheld the general validity of urban redevelopment statutes in the landmark case, *Berman v. Parker*.

In 1956, the Federal-Aid Highway Act gave state and federal government complete control over new highways, and often they were routed directly through vibrant urban neighborhoods—isolating or destroying many—since the focus of the program was to bring traffic in and out of the central cores of cities as expeditiously as possible and nine out of every ten dollars spent came from the federal government. This resulted in a serious degradation of the tax bases of many cities, isolated entire neighborhoods and meant that existing commercial districts were bypassed by the majority of commuters. Segregation continued to increase as communities were displaced and many African Americans and Latinos were left with no other option than moving into public housing while whites moved to the suburbs in ever-greater numbers.

In Boston, one of the country's oldest cities, almost a third of the old city was demolished—including the historic West End—to make way for a new highway, low-and moderate-income high-rises (which eventually became luxury housing), and new government and commercial buildings. This came to be seen as a tragedy by many residents and urban planners, and one of the centerpieces of the redevelopment—Government Center—is still considered an example of the excesses of urban renewal.

Urban Renewal and Nuclear Fallout Shelters

In the early 1960s, The Kennedy Administration worked with developer Louis Lesser to develop Barrington Plaza in Los Angeles, at the time the largest urban renewal project in the western United States, which also served as a nuclear fallout shelter during the peak of the Kennedy Administration's nuclear crisis.

Redlining and Segregation

Redlining began with the National Housing Act of 1934 which established the Federal Housing Administration (FHA) to improve housing conditions and standards, and later led to the formation of the Department of Housing and Urban Development (HUD). While it was designed to develop housing for poor residents of urban areas, that act also required cities to target specific areas and neighborhoods for different racial groups, and certain areas of cities were not eligible to receive loans at all. This meant that ethnic minorities could only secure mortgages in certain areas, and resulted in a large increase in the residential racial segregation in the United States.

This was followed by the Housing Act of 1937, which created the U.S. Housing Agency and the nation's first public housing program—the Low Rent Public Housing Program. This program began the large public housing projects that later became one of the hallmarks of urban renewal in the United States: it provided funding to local governments to build new public housing, but required that slum housing be demolished prior to any construction.

Reactions Against Urban Renewal

In 1961, Jane Jacobs published *The Death and Life of Great American Cities*, one of the first—and strongest—critiques of contemporary large-scale urban renewal. However, it would still be a few years before organized movements began to oppose urban renewal.

In 1964, the Civil Rights Act removed racial deed restrictions on housing. This began desegregation of residential neighborhoods, but redlining continued to mean that real estate agents continued to steer ethnic minorities to certain areas. The riots that swept cities across the country from 1965 to 1967 damaged or destroyed additional areas of major cities—most drastically in Detroit during the 12th Street Riot.

By the 1970s many major cities developed opposition to the sweeping urban-renewal plans for their cities. In Boston, community activists halted construction of the proposed Southwest Expressway—but only after a three-mile long stretch of land had been cleared. In San Francisco, Joseph Alioto was the first mayor to publicly repudiate the policy of urban renewal, and with the backing of community groups, forced the state to end construction of highways through the heart of the city. Atlanta lost over 60,000 people between 1960 and 1970 because of urban renewal and expressway construction, but a downtown building boom turned the city into the showcase of the New South in the 1970s and 1980s. In the early 1970s in Toronto Jacobs was heavily involved in a group which halted the construction of the Spadina Expressway and altered transport policy in that city.

From "Urban Renewal" to "Community Development"

Some of the policies around urban renewal began to change under President Lyndon Johnson and the War on Poverty, and in 1968, the Housing and Urban Development Act and The New Communities Act of 1968 guaranteed private financing for private entrepreneurs to plan and develop new communities. Subsequently, the Housing and Community Development Act of 1974 established the Community Development Block Grant program (CDBG) which began in earnest the focus on redevelopment of existing neighborhoods and properties, rather than demolition of substandard housing and economically depressed areas.

Currently, a mix of renovation, selective demolition, commercial development, and tax incentives is most often used to revitalize urban neighborhoods. An example of an entire eradication of a community is Africville in Halifax, Nova Scotia. Though not without its critics—gentrification is still controversial, and often results in familiar patterns of poorer residents being priced out of urban areas into suburbs or more depressed areas of cities—urban renewal in its present form is generally regarded as a great improvement over the policies of the middle part of the 20th century. Some programs, such as that administered by Fresh Ministries and Operation New Hope in Jacksonville, Florida attempt to develop communities, while at the same time combining highly favourable loan programs with financial literacy education so that poorer residents may still be able to afford their restored neighborhoods.

Urban Renewal Around the World

The Josefov neighbourhood, or Old Jewish Quarter, in Prague was levelled and rebuilt in an effort at urban renewal between 1890 and 1913. Other programs, such as that in Castleford in the UK and known as The Castleford Project seek to establish a process of urban renewal which enables local citizens to have greater control and ownership of the direction of their community and the way in which it overcomes market failure. This supports important themes in urban renewal today, such as participation, sustainability and trust – and government acting as advocate and 'enabler', rather than an instrument of command and control.

During the 1990s the concept of culture-led regeneration gained ground. Examples most often cited as successes include Temple Bar in Dublin where tourism was attracted to a bohemian 'cultural quarter', Barcelona where the 1992 Olympics provided a catalyst for infrastructure improvements and the redevelopment of the water front area, and Bilbao where the building of a new art museum was the focus for a new business district around the city's derelict dock area. The approach has become very popular in the UK due to the availability of lottery funding for capital projects and the vibrancy of the cultural and creative sectors.

However, while the arrival of Tate Modern in the London borough of Southwark may be heralded as a catalyst to economic revival in its surrounding neighbourhood, some civic authorities in the UK – for instance Newcastle-upon-Tyne and Gateshead have been accused of investing in cultural facilities at the cost of other programs and projects.

In post-apartheid South Africa major grassroots social movements such as the Western Cape Anti-Eviction Campaign and Abahlali base Mjondolo emerged to contest 'urban renewal' programs that forcibly relocated the poor out of the cities.

Long-Term Implications

While urban renewal has rarely lived up to the hopes of its original proponents – and has been hotly debated by politicians, urban planners, civic leaders, and residents of areas marked for renewal – it has played an undeniably important role.

Additionally, urban renewal can have many positive effects. Replenished housing stock might be an improvement in quality; it may increase density and reduce sprawl; it might have economic benefits and improve the global economic competitiveness of a city's centre. It may, in some instances, improve cultural and social amenity, and it may also improve opportunities for safety and surveillance. Developments such as London Docklands increased tax revenues for government. In late 1964 the British commentator Neil Wates expressed the opinion that urban renewal in the USA had 'demonstrated the tremendous advantages which flow from an urban renewal programme,' such as remedying the 'personal problems' of the poor, creation or renovation of housing stock, educational and cultural 'opportunities'.

As many examples listed above show, urban renewal has also been responsible for the destruction of existing communities; lead to social exclusion; loss of amenity; pollution; and displacement. Replacement housing – particularly in the form of housing towers – might be difficult to police, leading to an increase in crime, and such structures might in themselves be dehumanising. Urban renewal is usually non-consultative.

Urban renewal continues to evolve as successes and failures are examined and new models of development and redevelopment are tested and implemented.

Transport

Very densely built-up areas require high capacity urban transit, and urban planners must consider these factors in long term plans (Canary Wharf tube station). Although an important factor, there is a complex relationship between urban densities and car use. Transport within urbanized areas presents unique problems. The density of an urban environment increases traffic, which can harm businesses and increase pollution unless properly managed. Parking space for private vehicles requires the construction of large parking garages in high density areas. This space could often be more valuable for other development.

Good planning uses transit oriented development, which attempts to place higher densities of jobs or residents near high-volume transportation. For example, some cities permit commerce and multi-story apartment buildings only within one block of train stations and multilane boulevards, and accept single-family dwellings and parks farther away. Floor area ratio is often used to measure density. This is the floor area of buildings divided by the land area. Ratios below 1.5 are low density. Ratios above five very high density. Most exurbs are below two, while most city centres are well above five. Walk-up apartments with basement garages can easily achieve a density of three. Skyscrapers easily achieve densities of thirty or more.

City authorities may try to encourage higher densities to reduce per-capita infrastructure costs. In the UK, recent years have seen a concerted effort to increase the density of residential development in order to better achieve sustainable development. Increasing development density has the advantage of making mass transport systems, district heating and other community facilities (schools, health centres, etc.) more viable. However critics of this approach dub the densification of development as 'town cramming' and claim that it lowers quality of life and restricts market-led choice.

Problems can often occur at residential densities between about two and five. These densities can cause traffic jams for automobiles, yet are too low to be commercially served by trains or light rail systems. The conventional solution is to use buses, but these and light rail systems may fail where automobiles and excess road network capacity are both available, achieving less than 2% ridership. The

Lewis-Mogridge Position claims that increasing road space is not an effective way of relieving traffic jams as latent or induced demand invariably emerges to restore a socially-tolerable level of congestion.

Sub-urbanization

Sub-urbanization (or sub-urbanisation) is a term used to describe the growth of areas on the fringes of major cities. It is one of the many causes of the increase in urban sprawl.

Many residents of metropolitan areas no longer live and work within the central urban area, choosing instead to live in satellite communities called suburbs and commute to work via automobile or mass transit. Others have taken advantage of technological advances to work from their homes, and chose to do so in an environment they consider more pleasant than the city. These processes often occur in more economically developed countries, especially in the United States, which is believed to be the first country in which the majority of the population lives in the suburbs, rather than in the cities or in rural areas. Proponents of containing urban sprawl argue that sprawl leads to urban decay and a concentration of lower income residents in the inner city.

Causes and Effects

Sub-urbanization can be linked to a number of different push and pull factors. Push factors include the congestion and population density of the cities, pollution caused by industry and high levels of traffic and a general perception of a lower quality of life in inner city areas. Pull factors include more open spaces and a perception of being closer to "nature", lower suburban house prices and property taxes in comparison to the city, and the increasing number of job opportunities in the suburban areas.

Improvements in transportation infrastructure encourage Sub-urbanization, as people become increasingly able to live in a suburb and commute in to the nearby town or city to work. Developments in railways, bus routes and roads are the main improvements that make Sub-urbanization more practical. The increase in the number and size of highways is a particularly significant part of this effect.

Government policies can have a significant effect on the process. In the United States, for instance, policies of the Federal government in the post-World War II era, such as the building of an efficient network of roads, highways and super highways, and the underwriting of mortgages for suburban one-family homes, had an enormous influence on the pace of Sub-urbanization in that country. In effect, the government was encouraging the transfer of the middle-class population out of the inner cities and into the suburbs, sometimes with devastating effects on the viability of the city centres. However, some argue that the effect of Interstate Highway Systems on Sub-urbanization is overstated. Researchers of this vein believe city center populations would have declined even in the absence of highway systems, contending that Sub-urbanization is a long-standing and almost universal process. They primarily argue that as incomes rise, most people want the range and choice offered by automobiles. In addition, there is no significant evidence directly linking the development of highway systems to declining urban populations.

Insurance companies also fuelled the push out of cities, as in many cases, it redlined inner-city neighborhoods, denying mortgage loans there, and instead offering low rates in the suburban areas. More recently, some urban areas have adopted "green belt" policies which limit growth in the fringe of a city, in order to encourage more growth in the urban core. It began to be realized that a certain amount of population density in the center city is conducive to creating a good, working urban environment.

Race also played a role in American Sub-urbanization. During World War I, the massive migration of African Americans from the South resulted in an even greater residential shift toward suburban areas. The cities became seen as dangerous, crime-infested areas, while the suburbs were seen as safe places to live and raise a family, leading to a social trend known in some parts of the world as white flight. This phenomenon runs counter to much of the rest of the world, where slums mostly exist outside the city, rather than within them. With the increasing population of the older, more established suburban areas, many of the problems which were once seen as purely urban ones have manifested themselves there as well. Some social scientists suggest that the historical processes of Sub-urbanization and decentralization are instances of white privilege that have contributed to contemporary patterns of environmental racism.

Recent developments in communication technology, such as the spread of broadband services, the growth of e-mail and the advent of practical home video conferencing, has enabled more people to work from home rather than commuting. Although this can occur either in the city or in the suburbs, the effect is generally decentralizing, which works against the largest advantage of the center city, which is easier access to information and supplies due to centralization. Similarly, the rise of efficient package express delivery systems, such as (in the United States) Federal Express and UPS, which take advantage of computerization and the availability of an efficient air transportation system, also eliminates some of the advantages that were once to be had from having a business located in the city.

Industrial, warehousing, and factory land uses have also moved to suburban areas. Cheap telecommunications removes the need for company headquarters to be within quick courier distance of the warehouses and ports. Urban areas suffer from traffic congestion, which creates costs in extra driver costs for the company which can be reduced if they were in a suburban area near a highway. As with residential, lower property taxes and low land prices encourage selling industrial land for profitable brownfield redevelopment. Suburban areas also offer more land to use as a buffer between industrial and residential and retail space to avoid NIMBY sentiments and gentrification pressure from the local community when residential and retail is adjacent to industrial space in an urban area. Suburban municipalities can offer tax breaks, specialized zoning, and regulatory incentives to attract industrial land users to their area, such as City of Industry, California.

The overall effect of these developments is that businesses as well, and not just individuals, now see an advantage to locating in the suburbs, where the cost of buying land, renting space, and running their operations, is cheaper than in the city.

This has led to another recent phenomenon in American suburbs, the advent of edge cities in suburban areas, arising out of clusters of office buildings built around commercial strips and shopping malls.

With more and more jobs for suburbanites being located in these areas rather than in the main city core that the suburbs grew out of, traffic patterns, which for decades centered on people commuting into the center city to work in the morning and then returning home in the evening, have become more complex, with the volume of intra-suburban traffic increasing tremendously. By 2000, half of the US population lived in suburban areas.

11

Folk Society

Folk society, an ideal type or concept of society that is completely cohesive—morally, religiously, politically, and socially—because of the small numbers and isolated state of the people, because of the relatively unmediated personal quality of social interaction, and because the entire world of experience is permeated with religious meaning, the understanding and expression of which are shared by all members. The folk society is generally assumed to be the model of preliterate or so-called primitive societies that anthropologists have traditionally studied. The most important and enduring modern effort to make the concept of folk culture relevant to anthropology remains the work of the U.S. anthropologist Robert Redfield, who saw folk society as including not only primitive groups but also peasant peoples whose operations entailed some degree of dependence on the city. Although criticized for this interpretation of peasant life, as well as for underrating the impersonal and economic values and relations that may obtain in folk societies, Redfield's construction of the ideal folk culture continues to be the authoritative ideal type. Especially significant characteristics of folk society, as Redfield saw it, are its self-conception as the vessel of the sacred (this conception endowing the moral order with absolute authority and rendering the life-styles rigidly conventionalized) and its quality of being the whole of social and spiritual reality, with functions satisfying all the needs of an individual from birth, through all his life crises and transitions, to death.

Banjara

The Banjara are a class of usually ascribed as nomadic people from the Indian state of Rajasthan, North-West Gujarat, and Western

Madhya Pradesh and Eastern Sindh province of pre-independence Pakistan. They claim to belong to the clan of Agnivanshi Rajputs, and are also known as *Banjari, Pindari, Bangala, Banjori, Banjuri, Brinjari, Lamani, Lamadi, Lambani, Labhani, Lambara, Lavani, Lemadi, Lumadale, Labhani Muka, Goola, Gurmarti, Gormati, Kora, Sugali, Sukali, Tanda, Vanjari, Vanzara,and Wanji* Together with the Domba, they are sometimes called the "Gypsies of India". They are divided in three tribes, Maturia, Labana, Charan.]

The Banjara have spread to Andhra Pradesh, Karnataka, Maharashtra, Madhya Pradesh, Rajasthan, Uttar Pradesh and other states of India. About half their number speak Lambadi, one of the Rajasthani dialects of Hindustani, while others are native speakers of Hindi, Telugu and other languages dominant in their respective areas of settlement.

Rathore, Parmar, Pawar, Chauhan, castes belong to Vanjara community in Rajasthan and Gujarat now are in General Seats after the communal rights taken place in Rajasthan for Reservation in 2008 as they were landlords in Amarkot, Fathaykot and Sialkot before Partition of India and Pakistan While they are a in OBC in Andhra Pradesh (where they are listed as Sugali) and Orissa, a Scheduled Caste in Karnataka, Haryana, Punjab, and Himachal Pradesh.

The word *Banjara* is a deprecated, colloquial form of the word of Sanskrit origin. The sanskit bi-word *Vana chara* transliterated as "Forest wanderers," presumably because of their primitive role in the Indian society as forest wood collectors and distributors.

Culture

Food

The traditional food of Lambadis is Bati which is Roti. Daliya is a dish cooked using many cereal [wheat, jawar]. Banjara people are very much fascinated about non-vegetarian food. SALOI (made from goat blood and other parts of goat) is a non-vegetarian dish made exclusively by Banjara people. They prefer eating spicy food.

Dress

Women are known to wear colorful and beautiful costumes like PHETIYA [as Ghagra] and KANCHALLI [as top] and have tattoos on their hands. The dress is considered fancy and attractive by Western cultures. They use mirror chips and often coins to decorate it. Women put on thick bangles on their arms [PATLI]. Their ornaments are made up of silver rings, coins, chain and hair pleats are tied together at the end by CHOTLA. Men wear Dhoti and Kurta [short with many folds]. These clothes were designed specially for the protection from harsh climate in deserts and to distinguish them from others. These are the original inhabitants of Indus Valley Civilization who had to migrate to desert due to Islamic invasions from Himalayas and Hindu Kush.

Arts, Literature and Entertainment

Their customs, language and dress indicate they originated from Rajasthan. They live in settlements called thandas. They lived in zupada [hut]. Now many of them live in cities. They have a unique culture and dance form. On many occasions they gather, sing and dance. Their traditional occupation is agriculture and trade. Banjaras are also a group of nomadic cattle herders. The accurate history of Lambanis or Lambadis or Banjaras is not known but the general opinion among them is that they fought for Prithvi Raj against Muhammad of Ghor. The trail of the Lambadi/Banjara can be verified from their language, Lambadi borrows words from Rajasthani, Gujarati, Marathi and the local language of the area they belong to.

Banjaras originally belong to Rajasthan and they were Rajputs who migrated to southern parts of India for trade and agriculture. They settled down in the southern or central area of the country and slowly loosened contacts with Rajasthan, and their original community. Over a period of time both the communities separated and they adopted the local culture. The language spoken by Banjaras settled in Yavatmal district of Vidarbha, Maharashtra is an admixture of Hindi, Rajasthani and Marathi. The word "Banjara" must have evolved from Prakrit and Hindi and Rajasthani words "Bana/Ban or Vana/ Van" meaning Forest or Moorlands and "Chara" meaning 'Movers'. The Banjara are (together with the Domba) sometimes called the "Gypsies of India".

Dance, Lambadi is a special kind of dance of Andhra Pradesh. In this form of dance, mainly the female dancers dance in tune with the male drummers to offer homage to their Lord for a good harvest. At Anupu Village near Nagarjunakonda, Lambadi dance originated. They are actually semi-nomadic tribes who are gradually moving towards civilization. This dance is mainly restricted among the females and rarely the males participate in Lambadi Dance. Lambadi is a special kind of Folk Dance which involves participation by tribal women who bedeck themselves in colorful costumes and jewelry.

Related Communities

In Uttar Pradesh and Gujarat, there are several communities of Muslim Banjaras, who simply Muslim converts from the Banjara caste. The Muker, another Muslim community also traces its ancestry from the Banjara. Two other castes that claim kinship with the Banjara are the Labana of Punjab and Lavana of Rajasthan.

Indian Folk Music

Indian folk music is diverse because of India's vast cultural diversity. It has many forms including bhangra, lavani, dandiya and Rajasthani. The arrival of movies and pop music weakened folk music's popularity, but cheaply recordable music has made it easier to find and helped revive the traditions. Folk music (*desi*) has been influential on classical music, which is viewed as a higher art form. Instruments and styles have influenced classical ragas. It is also not uncommon for major writers, saints and poets to have large musical libraries and traditions to their name, often sung in *thumri* (semi-classical) style. Most of the folk music of India is dance-oriented.

Bhavageete

Bhavageete (literally 'emotion poetry') is a form of expressionist poetry and light music. Most of the poetry sung in this genre pertain to subjects like love, nature, philosophy etc, and the genre itself is not much different from Ghazals, though ghazals are bound to a peculiar metre. This genre is quite popular in many parts of India, notably in Karnataka. This genre may be called by different names in other languages. Kannada Bhavageete draws from the poetry of modern, including Kuvempu, D.R. Bendre, Gopalakrishna Adiga, K.S. Narasimhaswamy, G.S. Shivarudrappa, K. S. Nissar Ahmed, N S Lakshminarayana Bhatta etc. Notable Bhavageete performers include P. Kalinga Rao, Mysore Ananthaswamy, C. Aswath, Shimoga Subbanna, Archana Udupa, Raju Ananthaswamy etc.

Bhangra

Bhangra is a form of dance-oriented folk music that has become a pop sensation in the United Kingdom. The present musical style is derived from the traditional musical accompaniment to the folk dance of Punjab called by the same name, *bhangra* the female dance of punjab is known as gidda.

Bhangra is a form of dance and music that originated in the Punjab region. *Bhangra* dance began as a folk dance conducted by

Punjabi Sikh farmers to celebrate the coming of the harvest season. The specific moves of *Bhangra* reflect the manner in which villagers farmed their land. This dance art further became synthesized after the partition of India, when refugees from different parts of the Punjab shared their folk dances with individuals who resided in the regions they settled in. This hybrid dance became *Bhangra*. The folk dance has been popularised in the western world by Punjabi Sikhs and is seen in the West as an expression of South Asian culture as a whole. Today, *Bhangra* dance survives in different forms and styles all over the globe – including pop music, film soundtracks, collegiate competitions and even talent shows.

History

Bhangra dance is based on a Punjabi folk dhol beat called 'bhangra' singing and the beat of the dhol drum, a single-stringed instrument called the iktar (ektara), the tumbi and the chimta. Bhangra music however, is a form of music that originated in 1980s in Britain. The accompanying songs are small couplets written in the Punjabi language called *bolis*. They relate to current issues faced by the singers and (dil di gal) what they truly want to say. In Punjabi folk music, the dhol's smaller cousin, the dholki, was nearly always used to provide the main beat. Nowadays the dhol is used more frequently in folk music however in bhangra dholki is still preferred, with and without the dholki. Additional percussion, including tabla, is less frequently used in bhangra as a solo instrument but is sometimes used to accompany the dhol and dholki. The dholki drum patterns in Bhangra music bear an intimate similarity to the rhythms in Reggae music. This rhythm serves as a common thread which allows for easy commingling between Punjabi folk and Reggae as demonstrated by such artists as the UK's Apache Indian.

In the late 1960s and 1970s, several Punjabi Sikh bands from the United Kingdom set the stage for Bhangra to become a form of music instead of being just a dance. The success of many Punjabi artists based in the United Kingdom, created a fanbase, inspired new artists, and found large amounts of support in both East and West Punjab. These artists, some of whom are still active today, include, Heera Group, Alaap band, A.S. Kang and Apna Sangeet.

In the 1980s

Major migrations of the Sikh Punjabis to the UK brought with them the Bhangra music, which became popular in Britain during the 1980s, although heavily influenced in Britain by the infusion of rock sounds and a need to move away from the simple and repetitive punjabi folk music. It signaled the development of a self-conscious and distinctively British Asian youth culture centred on an experiential sense of self i.e. language, gesture, bodily signification, desires, etc... in a situation in which tensions with British culture and racist elements in British society had resulted in alienation in many minority ethnic groups and fostered a sense of need for an affirmation of a positive identity and culture, and provided a platform for British Asian males to assert their masculinity.

Bhangra dancing was originally perceived as a male dance, a "man's song", with strong, intense movements. However, "Second-generation South Asian American women are increasingly turning to bhangra as a way of defining cultural identity." In the 1980s Bhangra artists were selling over 30,000 cassettes a week in the UK, but not one artist made their way into the Top 40 UK Chart, despite these artists outselling popular British ones, as most sales were not through the large UK record stores whose sales were recorded by the Official UK Charts Company.

The 1980s is also what is commonly known as the golden age or what the "bhangraheads" refer to as the age of bhangra music which lasted roughly from 1985 to 1993. The primary emphasis during these times was on the melody/riff (played out usually on a synthesizer/harmonium/accordion or a guitar); the musician/composer received as much fanfare if not more, than the vocalist. The folk instruments were rarely used because it was agreed that the music was independent of the instruments being used. This era saw the very first boy band called the Sahotas, a band made up of five brothers from Wolverhampton, UK. Their music is a fusion: Bhangra, rock and dance fused with their very own distinctive sound.

One of the biggest Bhangra stars of the last several decades is Malkit Singh — known as "the golden voice of the Punjab" — and his group, Golden Star. Malkit was born in June 1963, in the village of Hussainpur in Punjab. He attended the Khalsa College, Jalandhar, in Punjab, in 1980 to study for a bachelor of arts degree. There he met his mentor, Professor Inderjit Singh, who nurtured his skills in Punjabi folk singing and Bhangra dancing. Due to Singh's tutelage, Malkit entered and won many song contests during this time. In 1983 he won a gold medal at the Guru Nanak Dev University, in Amritsar, Punjab, for performing his hit song "Gurh Naloo Ishq Mitha", which later featured on his first album, *Nach Gidhe Wich,* released in 1984. The album was a strong hit among South Asians worldwide, and after its release Malkit and his band moved to the United Kingdom to continue their work. Malkit has now produced 16 albums and has toured 27 countries in his Bhangra career. Malkit has been awarded the prestigious MBE by the British Queen for his services to Bhangra music. The group Alaap, fronted by Channi Singh, the man made famous by his white scarf, hails from Southall, a Punjabi area in London. Their album *Teri Chunni De Sitaray*, released in 1982 by Multitone, created quite a stir at a time when Bhangra was still in its early days in the UK. This album played a critical role in creating an interest in Bhangra among Asian university students in Britain. Alaap were unique with a live-set that was the best ever to play on the Bhangra stage. Their music oozed perfection, especially within the melody section. The music produced for Alaap included the pioneering sounds by Deepak Khazanchi.

Heera, formed by Bhupinder Bhindi and fronted by Kumar and Dhami, was one of the most popular bands of the 1980s. Fans were known to gate-crash weddings where they played. The group established itself with the albums *Jag Wala Mela*, produced by music maestro of the time Kuljit Bhamra and *Diamonds from Heera*, produced by Deepak Khazanchi, the man behind the new sound of UK Bhangra, on Arishma records. These albums are notable for being amongst the first Bhangra albums to successfully create mix Western drums and synthesizers with traditional Punjabi instruments.

Bands such as "Alaap" and "Heera" incorporated rock-influenced beats into Bhangra because it enabled "Asian youth to affirm their identities positively" within the broader environment of alternative Rock as an alternative way of expression. However, some believe that the progression of Bhangra music created an "intermezzo culture" post-India's Partition, within the unitary definitions of Southeast Asians within the diaspora, thus "establishing a brand new community in their home away from home".

Several other influential groups appeared around the same time, including The Saathies, Bhujungy Group, and Apna Sangeet. Apna Sangeet, most famously known for their hit "Mera Yaar Vajavey Dhol", re-formed in May 2009 after a break-up for charity. They are known as one of the best live acts in Bhangra.

When bhangra and Indian sounds and lyrics were brought together, British-Asian artists began incorporating them in their music. Certain Asian artists, such as Bally Sagoo, Talvin Singh, Badmarsh, Black Star Liner, and State of Bengal are creating their own form of British hip-hop. Even more well established groups like Cornershop, Fun-Da-Mental, and Asian Dub Foundation are finding different means and methods to create new sounds that other Asian groups have never formed. By mixing the sounds of bhangra with the popular sounds of hard rock and heavy metal, Asians are able to stay true to their own culture, while being open to a world of change. British Asians have to be conscious of both cultures in their everyday life and now are doing so in their music as well.

In the 1990s

Bhangra took large steps toward mainstream credibility in the 1990s, especially among youths. At the beginning of the nineties, many artists returned to the original, folk beats away from bhangra music, often incorporating more dhol drum beats and tumbi. This time also saw the rise of several young Punjabi singers. Beginning around 1994, there was a trend towards the use of samples (often sampled from mainstream hip hop) mixed with traditional folk rhythm instruments such as tumbi and dhol. Using folk instruments, hip hop samples, along with relatively inexpensive folk vocals imported from Punjab, Punjabi folk music was able to abolish Bhangra music.

An influential singer was the "Canadian folkster", Jazzy B. Originally from Namasher in Punjab, "Jaswinder Bains", as he is commonly referred to, his debut was in 1992. Having sold over 55,000 copies of his third album, *Folk and Funky*, he is now one of the best-selling Punjabi folk artist in the world, with a vocal style likened to that of Kuldip Manak. Although much of his music has a traditional Punjabi folk beat, he is known for having songs that incorporate a hip hop style such as "Romeo". Jazzy Bains gives wide recognition to the success of his many hits to Sukshinder Shinda, who has produced his music.

In the late 1990s Benjie Shah emerged as the heir to the Bhangra throne. His acoustic sound and his silky smooth voice made him an international sensation. He introduced many people to bhangra and his songs are still on the lips of his many fans. His UK #1 hit song "Short-handed" still holds many retail sales records.

Other influential folk artists include Surinder Shinda-famous for his "Putt Jattan De"-Harbhajan Mann, Manmohan Waris, Meshi Eshara, Sarbjit Cheema, Hans Raj Hans, Sardool Sikander, Sahotas, Geet the MegaBand, Anakhi, Sat Rang, XLNC, B21, Shaktee, Intermix, Sahara, Paaras, PDM, Amar Group, Sangeet Group, and Bombay Talkie. A dj to rise to stardom with many successful hits was Panjabi MC.

In 2010, the story of how Bhangra arrived and developed in the UK was told in the critically acclaimed stage musical *Britain's Got Bhangra*. Produced by Rifco Arts, and starring Shin (a genuine Bhangra star from band DCS), the show's story starts in Punjab in 1977 when a teenager (Twinkle) comes to the UK and, after finding success singing at Temples and weddings, charts the rise and fall of his career and Bhangra music right up to the present day. The show was the first ever Bhangra musical, with much of the show in Punjabi, making use of the live dhol, dholak and tabla throughout, and has traditional folk melodies as well as an original score by Sumeet Chopra with crossovers into pop, rap and R&B. The production had its world premiere at Theatre Royal Stratford East in May 2010 and went on a UK Tour. Another major tour is expected in 2011.

Dances

Bhangra has developed as a combination of dances from different parts of the Punjab region. The term "Bhangra" now refers to several kinds of dances and arts, including Jhumar, Luddi, Giddha, Julli, Daankara, Dhamal, Saami, Kikli, and Gatka.

- Jhumar, originally from Sandalbar, Punjab, comprises an important part of Punjab folk heritage. It is a graceful dance, based on a specific Jhumar rhythm. Dancers circle around a drum player while singing a soft chorus.
- A person performing the Luddi dance places one hand behind his head and the other in front of his face, while swaying his head and arms. He typically wears a plain loose shirt and sways in a snake-like manner. Like a Jhumar dancer, the Luddi dancer moves around a dhol player.
- Women have a different and much milder dance called Giddha. The dancers enact verses called bolis, representing a wide variety of subjects — everything from arguments with a sister-in-law to political affairs. The rhythm of the dance depends on the drums and the handclaps of the dancers.
- Daankara is a dance of celebration, typically performed at weddings. Two men, each holding colorful staves, dance around each other in a circle while tapping their sticks together in rhythm with the drums.
- Dancers also form a circle while performing Dhamal. They also hold their arms high, shake their shoulders and heads, and yell and scream. Dhamal is a true folk-dance, representing the heart of Bhangra.

- Women of the Sandalbar region traditionally are known for the Saami. The dancers dress in brightly colored kurtas and full flowing skirts called lehengas.
- Like Daankara, Kikli features pairs of dancers, this time women. The dancers cross their arms, hold each other's hands, and whirl around singing folk songs. Occasionally four girls join hands to perform this dance.
- Gatka is a Punjabi Sikh martial art in which people use swords, sticks, or daggers. Historians believe that the sixth Sikh guru started the art of Gatka after the martyrdom of fifth guru, Guru Arjan Dev. Wherever there is a large Punjabi Sikh population, there will be Gatka participants, often including small children and adults. These participants usually perform Gatka on special Punjabi holidays.

In addition to these different dances, a Bhangra performance typically contains many energetic stunts. The most popular stunt is called the moor, or peacock, in which a dancer sits on someone's shoulders, while another person hangs from his torso by his legs. Two-person towers, pyramids, and various spinning stunts are also popular.

Outfits

Traditional men wear a chaadra while doing Bhangra. A chaadra is a piece of cloth wrapped around the waist. Men also wear a kurta, which is a long Indian-style shirt. In addition, men wear pagadi (also known as turbans) to cover their heads. In modern times, men also wear turla, the fan attached to the pagadi. Colorful vests are worn above the kurta. Fumans (small balls attached to ropes) are worn on each arm. Women wear a traditional Punjabi dress known as a salwar kameez, long baggy pants tight at the ankle (salwar) and a long colorful shirt (kameez). Women also wear chunnis, colorful pieces of cloth wrapped around the neck.

These items are all very colorful and vibrant, representing the rich rural colors of Punjab. Besides the above, the Bhangra dress has different parts that are listed below in detail:

- Turla or Torla, which is a fan like adornment on the turban
- Pag (turban, a sign of pride/honor in Punjab). This is tied differently than the traditional turban one sees Sikhs wearing in the street. This turban has to be tied before each show
- Kurta-Similar to a silk shirt, with about 4 buttons, very loose with embroidered patterns.

- Lungi or Chadar, A loose loincloth tied around the dancer's waist, which is usually very decorated.
- Jugi: A waistcoat, with no buttons.
- Rumâl: Small 'scarves' worn on the fingers. They look very elegant and are effective when the hands move during the course of bhangra performance.

And you can see a photo of a bhangra dhol drummer, costumed and in full swing. According to Sanjay Sharma, in her article, she explains/points out the fact that Bhangra represents Asians and is referred to today as Asian music which accounts for the vast existence of Asian wear and not to mention symbols as part of their traditional dress/costumes.

Lyrics

Bhangra lyrics, always sung in the Punjabi language, generally cover social issues such as love, relationships, money, dancing, getting drunk and marriage. Additionally, there are countless Bhangra songs devoted to Punjabi pride themes and Punjabi heroes. The lyrics are tributes to the rich cultural traditions of the Punjabis. In particular, many Bhangra tracks have been written about Udham Singh and Bhagat Singh. Less serious topics include beautiful ladies with their colorful duppattas, and dancing and drinking in the fields of the Punjab.

Bhangra singers do not sing in the same tone of voice as their Southeast Asian counterparts. Rather, they employ a high, energetic tone of voice. Singing fiercely, and with great pride, they typically add nonsensical, random noises to their singing. Likewise, often people dancing to Bhangra will yell phrases such as *hoi, hoi, hoi*; *balle balle*; *chak de*; *oye hoi*; *bruah* (for an extended length of about 2–5 seconds); *haripa* or *ch-ch* (mostly used as slow beats called Chummer/Jhoomer) to the music. Some of the more famous Bhangra or Punjabi lyricists include Harbans Jandu (Jandu Littranwala) who has written famous songs like "*Giddhian Di Rani*".

Instruments

Many different Punjabi instruments contribute to the sound of Bhangra. Although the most important instrument is the keyboard, Bhangra also features a variety of string and other drum instruments.

The primary and most important instrument that defines Bhangra is the dhol. The dhol is a large, high-bass drum, played by beating it with two sticks-known as *daggah* (bass end) and *tilli* (treble end). The width of a dhol skin is about fifteen inches in general, and the dhol player holds his instrument with a strap around his neck. The string instruments include the guitar (both acoustic and electrical), bass, sitar, tumbi, violin and sarangi. The snare, toms, dhad, dafli, dholki, and damru are the other drums. The tumbi, originally played by folk artists such as Lalchand Yamla Jatt and Kuldip Manak in true folk recordings and then famously mastered by chamkila, a famous Punjabi folk singer (not bhangra singer), is a high-tone, single-string instrument. It has only one string, however it is difficult to master. The sarangi is a multi-stringed instrument, somewhat similar to the violin and is played using meends. The sapera produces a beautiful, high-pitched stringy beat, while the supp and chimta add an extra, light sound to Bhangra music. Finally, the dhad, dafli, dholki, and damru are instruments that produce more drum beats, but with much less bass than the dhol drum.

The keyboard and guitar are the most important melodic instruments used in bhangra with even the sitar being used on certain albums.

Percussion

With skilled keyboard, tumbi, guitar & dhol players, Bhangra today has evolved into a largely beat-based music genre, unlike until 1994 when it was slightly more mellow & classical. Pandit Dinesh and Kuljit Bhamra were trained exponents of Indian percussion and helped create the current UK sound, albeit mainly with tabla and dholki for bands like Alaap and Heera. The generation that followed became overly dependent on folk music.

A talented 15 year old percussionist called Bhupinder Singh Kullar, aka 'Tubsy' of Handsworth, Birmingham created a more contemporary style and groove that seemed to fuse more naturally with western music. Songs such as Dhola veh Dhola (Satrang) and albums such as Bomb the Tumbi (Safri Boyz) contained this new style and were very successful.

Then came Sunil Kalyan of Southall, London who also sessioned on many songs and albums. He added a smoothness and sweetness never heard before on the tabla, hailing him as one of the best tabla players in UK Bhangra.

Sukhshinder Shinda later introduced his unique style of dhol playing with the album 'Dhol Beat.' He added a very clean style of dhol playing and helped create the sound for artists such as Jaswinder Singh Bains and Bhinda Jatt. He was regarded at the time as the best Dhol player in UK.

Another influential percussionist was Parvinder Bharat (Parv) of Wolverhampton, who for many years had been percussionist for DCS, his style, speed and improvisational skills were second to none. Parv also introduced playing the Dholak and tabla top end (dhayan) with great effect into the live bhangra scene, a style that has been adopted by most bhangra percussionists ever since.

Other important percussionists include Juggy Rihal of Coventry, Aman Hayer and Billy Sandher of Gravesend.

Remixes

Punjabi folk remixed with hip hop, known lovingly as folkhop, is most often produced when folk vocals are purchased online to be remixed in a studio. Folk vocals are usually sung out to traditional melodies, that are often repeated with new lyrics. This genre is considered a sub genre of Punjabi folk and not accepted as bhangra music.

Many South Asian DJs, especially in America, have mixed Punjabi folk music with house, reggae, and hip-hop to add a different flavor to Punjabi folk. These remixes continued to gain popularity as the nineties came to an end.

Of particular note among remix artists is Bally Sagoo, a Punjabi-Sikh, Anglo-Indian raised in Birmingham, England. Sagoo described his music as "a bit of tablas, a bit of the Indian sound. But bring on the bass lines, bring on the funky-drummer beat, bring on the James Brown samples", to *Time* magazine in 1997. He was recently signed by Sony as the flagship artist for a new sound. The most popular of these is Daler Mehndi, a Punjabi singer from India, and his music, known as "folk Pop". Mehndi has become a major name not just in Punjab, but also all over India, with tracks such as "Bolo Ta Ra Ra" and "Ho Jayegee Balle Balle". He has made the sound of Bhangra-pop a craze amongst many non-Punjabis in India, selling many millions of albums. Perhaps his most impressive accomplishment is the selling of 250,000 albums in Kerala, a state in the South of India where Punjabi is not spoken.

Toward the end of the decade, Bhangra continued to die out, with folkhop artists like Bally Sagoo and Apache Indian signing with international recording labels Sony and Island. Moreover, Multitone Records, one of the major recording labels associated with Bhangra in Britain in the eighties and nineties, was bought by BMG. Finally, a recent Pepsi commercial launched in Britain featured South Asian actors and Punjabi folk music. This, perhaps more than anything else, is a true sign of the emergence of Punjabi folk into popular culture.

Post-Bhangra continues to gain popularity in both the UK and US after the death of bhangra in mid 90s. As mentioned above, artists such as Bally Sagoo offer what was referred to as "Bollywood remixes". This is just one result of the fusion the traditional folk beats and South Asian instruments with that of other contemporary music genres. Other lesser popular offshoots include "Bhangramuffin" and Acid Bhangra. Bhangramuffin mixes traditional Bhangra backgrounds are combined with Ragga; one famous band from this genre is Apache Indian. As the title suggests, Acid Bhangra combines acid music with Bhangra. An interesting result of its popularity was that post-Bhangra gave rise to a new wave of club culture (i.e. Hot 'n Spicy at London's Limelight nightclub and Manchester's Shankeys Soap). Although much of the popularity was centered on South Asian participation, post-Bhangra expressed a "process of musical cultural hybridization and syncretism that moved beyond a straightforward juxtaposition of dance music genres."

Although it sounds like a musical form that would follow the original Bhangra, post-Bhangra most specifically refers to a similar musical form with a greater emphasis on inter-dance-genre dialogues. In this way, post-Bhangra has an element of remixing and fusing Black and Asian styles of dance that is not as prevalent in traditional Bhangra. According to Sanjay Sharma, "just as Bhangra has been in constant dialogue with other (black) dance genres, post-Bhangra carries this through more incisively and intentionally." At the same time though, post-Bhangra contains the same components of racial and cultural affirmation that have been seen in Bhangra before it. In fact, with a larger focus on the dance fusion style and post-Bhangra "[operating] musically more in terms of other genres, of Ragga, Rap or Jungle music," it is very easy to see how this music attacks the essentialism that lay at the heart of the British Empire. Indeed, by fusing such starkly contrasting dance genres, post-Bhangra artists subject racial signifiers such as "Asianness" or "Blackness" to immense scrutiny. While these notions of racial and cultural essentialism can produce national pride and a finite sense of identity, they can also be highly detrimental to society at large.

Since essentialism claims that people can be categorized according to some definite essence, it often leads to civil disputes and factionalism and places social boundaries between different groups of people, preventing cultural diffusion.

On the flip side, post-Bhangra offers these displaced Asians in the UK an avenue for expressing their condemnation of the rigid essentialism which questions their participation in a black-dominated music scene in the first place. Not only this, but Asian artists address the issue that they too have faced social difficulties in the UK and that their music is truly authentic.

Post-bhangra has been described as a tool for strategic identity politics, presenting itself as a medium for protest against colonialism and racism. By raising awareness to problems of racism within Asian contexts, post-bhangra ventures to model movements created by organizations that have engendered a sense of unique identity for their constituents, such as the Nation of Islam and the Black Panthers. Furthermore, it allows for the expression of frustration related to racial tension that has been blanketed by social apathy. Post-bhangra music has created "new ethnicities" that are pertinent to the historical ties of the Indian subcontinent whilst remaining independent of the region itself. The formation of this new sense of Asian identity has been a leading proponent of post-bhangra, providing Asian youth with a feeling of belonging despite having to struggle to identify with a new culture that is not confined to the cultural norms of their roots nor completely assimilated with any existing culture.

Somewhere between mimicry and appropriation, post-Bhangra manages to feed off of mainstream black dance genres and then flourish in a more localized context, giving it the local importance it has to Asian and Black Britons. Rupa Huq says with respect to post-Bhangra's mainstream popularity that "spring 1998 saw the number one hit 'Brimful of Asha' from Cornershop," indicating that post-Bhangra is gaining the global attention it needs. As this musical form finds its way into mainstream styles and media, the British faces behind it hope to endorse the fight against "white racial terror, neo-colonialism/imperialism and global racial subjugation." Therefore, post-Bhangra is an important musical form both because it defies essentialism by mixing and clashing seemingly incommensurbale sounds and cultures and because it provides Asian youth in the UK with a vehicle for self-expression as part of a musical scene with which they would otherwise be discouraged from associating.

Cultural Impact

The interpretation of Bhangra must exist in the space where Asian, UK and hip hop cultures meet. There is an expressed concerned that oversimplification of the genre by outsiders is detrimental to the music's message, but artists are responsible for how they express their music's content as well. In "Bhangra's Ambassador, Keeping the Party

Spinning" from the New York Times, DJ Rekha is conscious of her cultural accountability to her music. She suggests that "because I'm working with my culture, and it's being accessed or consumed by other cultures, then Bhangra followers often feel that the music is an expression of identity. As the movement gains momentum, Bhangra music has also gained international recognition. "Asian fusion is a melding of the sounds of the sub-continent with hip-hop beats and R&B influences, and it's no longer destined to be tucked away in the World Music section of your record store." (Mehta, Ashta).

In North America

Punjabi immigrants have encouraged the growth of Punjabi folk music/rural music in the western hemisphere instead of bhangra music. The bhangra industry has not grown in North America nearly as much as it has grown in the United Kingdom. Indian Lion, a Canadian folk artist explains why:

The reasons there's a lot of bands in England is because there's a lot of work in England. In England the tradition that's been going on for years now is that there's weddings happening up and down the country every weekend, and it's part of the culture that they have Bhangra bands come and play, who get paid 1800 quid a shot, you know. Most of the bands are booked up for the next two years. And England is a country where you can wake up in the morning and by lunchtime you can be at the other end of the country, it helps. In Canada it takes 3 days to get to the other side of the country, so there's no circuit there. And it isn't a tradition [in Canada] to have live music at weddings. There are a few bands here that play a few gigs, but nothing major. —*Indian Lion*

However, with the emergence of North American (non bhangra) folk artists such as Manmohan Waris, Jazzy Bains, Kamal Heer, Harbhanjan Maan, Sarabjit Cheema, and Debi Makhsoospuri, and the growth of the remix market, the future of Punjabi folk music in North America looks good. Sanjay Sharma argues in "Noisy Asians or 'Asian Noise'?" that in the case of bhangra, musical exchanges have not been unidirectional but that in fact, a constant dialogue with other musical sources has taken place.

One of those musical sources is North America. More specifically, bhangra and reggae rhythms, stemming from Jamaica, have been easily fused. While black dance music beats have proved to mesh well with the fast-paced tempo of bhangra, Sharma notes reggae music also contains a *bhangara* rhythm related to the dhol drum. According to an article published on *SikhSpectrum.com*, the dhol drum has even been involved in the appropriation of bhangra in North American culture during Britain's neocolonial period. The arrival of bhangra in North America carries not only sonic importance, but also social and anthropological significance as it represents group cooperation and universal exchange between races, nations, and identities.

However, Sharma also notes the difficulty of bhangra's acceptance in the UK. Bhangra has been seen to be an 'otherness' of a British/ South Asian experience. It has also been seen that British culture mostly negates and derides this Asian expression. In 2001, the arch enemy of bhangra, Punjabi folk, and its hip hop form, folkhop began to exert an influence over US R&B music, when Missy Elliott released the folk hop-influenced song "Get Ur Freak On". In 2003, Punjabi MC's "Mundian To Bach Ke" (Beware of the Boys) was covered by the U.S. rapper Jay-Z.

The great popularity of these two tracks led to an even greater usage of Punjabi folk in American music. Additionally, American rapper Pras of The Fugees has recorded tracks with British alternative bhangra band Swami. Because the original Punjabi folk beat is different from the commercialized version we see today, the use of bhangra beats shows the complexity and ingenuity of hip-hop in North America and how artists gain inspiration from all different genres of music. The commercialization of Punjabi folk and the way it has traveled around the world speaks to the versatility and longevity of the musical style.

bhangra, the dance, has also expanded into the world of fitness. Fitness instructors like television host Sarina Jain have developed fitness routines based on bhangra dance moves for their workout programs.

Bhangra Dance Competitions

Bhangra competitions have been held in the Punjab for many decades. However, now universities and other organizations have begun to hold annual Bhangra dance competitions in many of the main cities of the United States, Canada, and England. At these competitions, young Punjabis, other South Asians, and people with no South Asian background compete for money and trophies. For example, Bruin Bhangra in Los Angeles has become one of the biggest bhangra competition in the nation. Teams from all over United States and Canada come together to compete and show their talent. Every year Bruin Bhangra also invites different well known Punjabi singers. SoCal Bhangra's past list of artists includes RDB, Manak-e,

Sukhshinder Shinda, Jassi Sidhu, KS Makhan and Malkit Singh. The eastcoast bhangra team of The George Washington University's South Asian Society hosts the biggest and most prestigious bhangra competition in the United States: Bhangra Blowout. With a crowd of many thousands, it serves as the signature culminating event of the year for the best college bhangra teams. Many regard it as a de facto national championship due to its large scale and the fact that it is the last significant bhangra competition of the American school year.

2010 was the first year for Elite 8 Bhangra Invitational, in Washington DC. This event invited 8 of the top teams from North America to showcase their routines and compete for the number one spot.

In the West, unlike the Punjab, there is less emphasis on traditional songs, and more focus on the flow of a mix; an easy example of this is a team such as "Da Real Punjabis", which is notorious for mixing traditional Bhangra music with hip hop or rock songs. This synergy of the Bhangra dance with other cultures` parallels the music's fusion with different genres. University competitions have experienced an explosion in popularity over the last five years and have helped to promote the dance and music in today's mainstream culture.

In the UK, the first ever major Bhangra Competition "The Bhangra Showdown" was organised by students from Imperial College London and held on 1 December 2007. The competition was held at Indigo2 in the O2 in Greenwich, and was attended by over 1000 people. All proceeds from this show were donated to two charities, Wateraid and The Child Welfare Trust, and the show looks to continue on an annual basis. The show was held once again on 31 January 2009 at the Sadler's Wells Theatre, with proceeds going to the MND Association and The Child Welfare Trust and was attended by around 1,500 people. Six universities took part: Imperial, Queen Mary's, Kingston, Brunel, Birmingham and Leicester/DMU. Birmingham came in 3rd place, Imperial came a very close 2nd and Queen Mary's took 1st place.

The largest student-run organization in Pittsburgh was a bhangra competition called Bhangra In The Burgh. Bhangra In The Burgh was organized by students from Carnegie Mellon University, and is held yearly, this year on January 30, 2010 at the Soldiers and Sailors Memorial in Oakland. Over 2100 people attended last year, and the show was sold out. All proceeds go to the Children's Home of Pittsburgh.

Relation to other Indian Dance Forms

Bhangra can be related to Assam's Bihu dance performed during Bihu festival. Magh Bihu is associated with farming; as the traditional

Assamese society is predominantly dependent on farming. Parallels between Bhangra Bihu can be drawn from the fact that the merriments for both of these music forms involve characteristic overtones with dances along with the enthralling beats of percussive musical instruments. Moreover, both Bihu and Bhangra involve high energy dance moves and sequences with young dancers in colorful clothing and the folk music played with the dhol. Both music/dance varieties, though having the similar themes, are distinctly different and have their proper origins in the respective regions of Punjab and Assam Valley.

Beat & Boom

Unique Radio Show on 104.9 FM Dallas/Ft Worth, TX. is devoted to new style of Bhangra. Show is broadcasted every Sunday from 7pm to 9pm. New artist can submit music on their website to be played on air. 104.9 FM in Dallas is one of the only 24/7 FM station in the USA that targets South Asian community living abroad.

Lavani

Lavani is a popular folk form of Maharashtra. Traditionally, the songs are sung by female artists, but male artists may occasionally sing Lavanis. The dance format associated with Lavani is known as Tamasha.

Dandiya

Dandiya is a form of dance-oriented folk music that has also been adapted for pop music worldwide, popular in Western India, especially during Navaratri. The present musical style is derived from the traditional musical accompaniment to the folk dance of Dandiya called by the same name, dandiya. Raas or Dandiya Raas is the traditional folk dance form of Vrindavan, India, where it is performed depicting scenes of Holi, and lila of Krishna and Radha. Along with Garba, it is the featured dance of Navratri evenings in Western India.

Etymology

The word "Raas" comes from Sanskrit word "Ras". The origins of Raas can be traced to ancient times. Lord Krishna performed Rasa lila" (Lila means Lord Krishna's playful dance. The word "Lila" also refers to things that God does that we do not fully understand).

Forms of Raas

There are several forms of Raas, but "Dandiya Raas", performed during Navaratri in Gujarat is the most popular form. Other forms of Raas include Dang Lila from Rajasthan where only one large stick is used, and "Rasa lila" from North India. Raas Lila and Dandiya Raas are similar. Some even consider "Garba" as a form of Raas, namely "Raas Garba".

In Dandiya Raas men and women dance in two circles, with sticks in their hands. In the old times Raas did not involve much singing, just the beat of Dhol was enough. "Dandiya" or sticks, are about 18" long. Each dancer holds two, although some times when they are short on Dandiya they will use just one in right hand. Generally, in a four beat rhythm, opposite sides hit the sticks at the same time, creating a nice sound. One circle goes clockwise and another counter clockwise. In the west, people don't form full circles, but instead often form rows.

Origin of Dandiya Raas

Originating as devotional Garba dances, which were always performed in Durga's honour, this dance form is actually the staging of a mock-fight between the Goddess and Mahishasura, the mighty demon-king, and is nicknamed "The Sword Dance". During the dance, dancers energetically whirl and move their feet and arms in a complicated, choreographed manner to the tune of the music with various rhythms. The dhol is used as well as complementary percussion instruments such as the dholak, tabla and others.

The sticks (*dandiya*s) of the dance represent the sword of Durga. The women wear traditional dresses such as colorful embroidered choli, ghagra and bandhani dupattas (traditional attire) dazzling with mirror work and heavy jewellery. The men wear special turbans and kedias, but this varies regionally.

Garba is performed before Aarti (worshipping ritual) as devotional performances in the honor of the Goddess, while Dandiya is performed after it, as a part of merriment. Men and women join in for Raas Dandiya, and also for the Garba.

The circular movements of Dandiya Raas are much more complex than those of Garba. The origin of these dance performances or Raas is Krishna. Today, Raas is not only an important part of Navratri in Gujarat, but extends itself to other festivals related to harvest and crops as well. The Mers of Saurastra are noted to perform Raas with extreme energy and vigor.

History

The Dandiya Raas dance originated as devotional Garba dances, which were performed in Goddess Durga's honor. This dance form is actually the staging of a mock-fight between Goddess Durga and Mahishasura, the mighty demon-king. This dance is also nicknamed 'The Sword Dance'. The sticks of the dance represent the sword of Goddess Durga. The origin of these dances can be traced back to the life of Lord Krishna. Today, Raas is not only an important part of Navaratri in Gujarat but extends itself to other festivals related to harvest and crops as well.

Format

Raas is also performed at social functions and on stage. Staged Raas can be very complex with intricate steps and music. Raas is a folk art and it will change with the times. When African slaves and ship workers (who were Muslims) arrived on the coast of Saurashtra, they adopted Raas as their own and used African drums. While it originated from Hindu tradition, it was adopted by the Muslim community as Saurashtra. Singing entered the Raas scene later on. Initially, most songs were about Lord Krishna, but songs about love, praise of warriors who fought gallant wars, and the Goddess Durga, and even Muslim Raas songs were born. It is common to think that Raas has to be fast, but that is not the case. Grace and slow movements are just as important.

With the advent of C-60 cassettes came the pre-recorded "non stop" Raas music. Soon it overtook the individual Raas items which are rarely recorded nowadays. The disco beat and use of western drum became popular, but you can still visit fine arts college in Vadodara during Navaratri where the musicians sit in the centre and play while people dance around them. Gujarati movies entered the scene in late 50's and 60's. Raas took on a different form as it borrowed heavily from the film industry.

There are other unique forms of Raas such as one in the town of Mahuva where men would tie one hand to a rope extending from above and hold a stick in the other hand. This was strictly in praise of Goddess Durga. If you use broader definition, even "Manjira" can be used to do Raas. There are communities that specialize in Raas with "Manjira". Just like the British police, some men dancing at "Tarnetar" used to wear colourful bands of cloth around their legs, resembling socks. The city of Mumbai developed its own style of Dandiya Raas. Now, during Navratri people use Dandiya, but make it more like a free style dance. "Head bobbing" during Raas is popular in USA among youngsters, but that arrived from the Gujarati movies. Head bobbing was for the singers, not for the dancers.

Costumes and Music

The women wear traditional dresses such as colorful embroidered choli, ghagra and bandahni dupattas, which is the traditional attire, dazzling with mirror work and heavy jewellery. The men wear special turbans and kedias, but can range from area to area. The dancers whirl and move their feet and arms in a choreographed manner to the tune of the music with a lot of drum beats. The dhol is used as well as complementary percussion instruments such as the dholak, tabla, etc. the true dance gets extremely complicated and energetic. Both of these dances are associated with the time of harvest.

Difference between Dandiya and Garba

The main difference between the Garba and Dandiya dance performances is that Garba is performed before Aarti (worshipping ritual) as devotional performances in honor of Goddess Durga while Dandiya is performed after it, as a part of merriment. While Garba is performed exclusively by women, men and women join in for Raas Dandiya. Also known as 'The Dance Of Swords' as performers use a pair of colorfully decorated sticks as symbols, the circular movements of Dandiya Raas are much more complex than that of Garba.

Amongst Indian diaspora

A new form of Raas is taking place in the USA. This is mostly a show item where college students of Indian origin mix non-stop Raas music with strong drum beats and stunts along with "themes" such as wedding, Star Wars and Lion King. They freely mix traditional steps with other steps. Raas will always be dynamic as it represents the circle of life, beating of heart. It is a live folk form that has changed with time and will keep changing.

Pandavani

Pandavani is a folk singing style of musical narration of tales from ancient epic Mahabharata with musical accompaniment and Bhima as hero. This form of folk theatre is popular in the Indian state of Chhattisgarh and in the neighbouring tribal areas of Orissa and Andhra Pradesh. Teejan Bai is most renowned singer to this style, followed by Ritu Verma..

Origins

The origins of this singing style are not known, and according to its foremost singer Teejan Bai, it might be as old as the Mahabharata itself, as few people could read in those times, and that is how perhaps they passed on their stories, generation after generation.

Overview

Pandavani, literally means stories or songs of Pandavas, the legendary brothers of Mahabharat, and involves the lead singer, enacting and singing with an ektara or a *tambura* (stringed musical instrument), decorated with small bells and peacock feathers in one hand and sometimes *kartal* (a pair of cymbals) on another.

It is part of the tradition of the tellers-of-tales present in every culture or tradition (like Baul singers of Bengal and Kathak performers), where ancient epics, anecdotes and stories are recounted, or re-enacted to educated and entertain the masses. Without the use of any stage props or settings, just by the use to mimicry and rousing theatrical movements, and in between the singer-narrator break into an impromptu dance, at the completion of an episode or to celebrate a victory with the story being retold, yet in its truest sense Pandavani remains an accomplished theatre form. During a performance, as the story builds, the tambura becomes a prop, sometimes it becomes to personify a *gada*, mace of Arjun, or at times his bow or a chariot, while others it becomes the hair of queen Draupadi or Dushshan thus helping the narrator-singer play all the characters of story.

The singer is usually supported by a group of performers on Harmonium, Tabla, Dholka, Majira and two or three singers who sing the refrain and provide backing vocals. Each singer adds his or her unique style to the singing, sometimes adding local words, improvising and offering critique on current happenings and an insights through the story. Gradually as the story progresses the performance becomes more intense and experiential with added dance movements, an element of surprise often used. The lead singer continuously interacts with the accompanying singers, who ask questions, give commetary, interject thus enhancing the dramatic effect of the performance, which can last for several hours on a single episode of Mahabharata. Eventually what starts out as a simple story narration turns into full-flegded ballad.

Variations (Shaili)

- Vedamati – the sitting style, mainly used by women, basically invented by Jhaduram Devangan.
- Kapalik – the traditional form, the standing style, where the performer depicts scenes from the epic and improvises consistently. (Teejan Bai's style)

Impact on Popular Culture

Influences of Pandanavi can been clearly seen in the plays of Habib Tanvir, who has been using folk singers of Chattisgarh in his plays, creating a free-style story narration format, typical of Pandavani.

Rajasthani

Rajasthani music has a diverse collection of musician castes, including langas, sapera, bhopa, jogi and Manganiar.

Bauls

The Bauls of Bengal were an order of musicians in 18th, 19th and early 20th century India who played a form of music using a khamak, ektara and dotara. The word Baul comes from Sanskrit *batul* meaning *divinely inspired insanity*. They are a group of Hindu mystic minstrels. They are thought to have been influenced greatly by the Hindu tantric sect of the Kartabhajas as well as by Sufi sects. Bauls travel in search of the internal ideal, *Maner Manush* (*Man of the Heart*).

Garba

"Garba (song), the songs sung in honor of Hindu goddesses during Navratri."

Dollu Kunita

This is a group dance that is named after the Dollu-the percussion instrument used in the dance. It is performed by the menfolk of the Kuruba community of the North Karnataka area. The group consists of 16 dancers who wear the drum and beat it to different rhythms while also dancing. The beat is controlled and directed by a leader with cymbals who is positioned in the center. Slow and fast rhythms alternate and group weaves varied patterns.

Kolata

Kolata is the traditional folk dance of the state of Karnataka, located in Southern India on the western coast. Similar to its North Indian counterpart Dandiya Ras, it is performed with coloured sticks and usually involves both men and women dancing together.

Veeragase

Veeragase is a dance folk form prevalent in the state of Karnataka, India. It is a vigorous dance based on Hindu mythology and involves very intense energy-sapping dance movements. Veeragase is one of the dances demonstrated in the Dasara procession held in Mysore. This dance is performed during festivals and mainly in the Hindu months of Shravana and Karthika.

12

The Village Community

Settlement and Structure

Scattered throughout India are approximately 500,000 villages. The Census of India regards most settlements of fewer than 5,000 as a village. These settlements range from tiny hamlets of thatched huts to larger settlements of tile-roofed stone and brick houses. Most villages are small; nearly 80 percent have fewer than 1,000 inhabitants, according to the 1991 census. Most are nucleated settlements, while others are more dispersed. It is in villages that India's most basic business—agriculture—takes place. Here, in the face of vicissitudes of all kinds, farmers follow time-tested as well as innovative methods of growing wheat, rice, lentils, vegetables, fruits, and many other crops in order to accomplish the challenging task of feeding themselves and the nation. Here, too, flourish many of India's most valued cultural forms.

Viewed from a distance, an Indian village may appear deceptively simple. A cluster of mud-plastered walls shaded by a few trees, set among a stretch of green or dun-colored fields, with a few people slowly coming or going, oxcarts creaking, cattle lowing, and birds singing—all present an image of harmonious simplicity. Indian city dwellers often refer nostalgically to "simple village life." City artists portray colorfully garbed village women gracefully carrying water pots on their heads, and writers describe isolated rural settlements unsullied by the complexities of modern urban civilization. Social scientists of the past wrote of Indian villages as virtually self-sufficient communities with few ties to the outside world. In actuality, Indian village life is far from simple. Each village is connected through a variety of crucial horizontal linkages with other villages and with urban areas both near and far. Most villages are characterized by a multiplicity of economic, caste, kinship, occupational, and even religious groups linked vertically within each settlement. Factionalism is a typical feature of village politics. In one of the first of the modern anthropological studies of Indian village life, anthropologist Oscar Lewis called this complexity "rural cosmopolitanism."

Throughout most of India, village dwellings are built very close to one another in a nucleated settlement, with small lanes for passage of people and sometimes carts. Village fields surround the settlement and are generally within easy walking distance. In hilly tracts of central, eastern, and far northern India, dwellings are more spread out, reflecting the nature of the topography. In the wet states of West Bengal and Kerala, houses are more dispersed; in some parts of Kerala, they are constructed in continuous lines, with divisions between villages not obvious to visitors.

In northern and central India, neighbourhood boundaries can be vague. The houses of Dalits are generally located in separate neighborhoods or on the outskirts of the nucleated settlement, but there are seldom distinct Dalit hamlets. By contrast, in the south, where socioeconomic contrasts and caste pollution observances tend to be stronger than in the north, Brahman homes may be set apart from those of non-Brahmans, and Dalit hamlets are set at a little distance from the homes of other castes.

The number of castes resident in a single village can vary widely, from one to more than forty. Typically, a village is dominated by one or a very few castes that essentially control the village land and on whose patronage members of weaker groups must rely. In the village of about 1,100 population near Delhi studied by Lewis in the 1950s, the Jat caste (the largest cultivating caste in northwestern India) comprised 60 percent of the residents and owned all of the village land, including the house sites. In Nimkhera, Madhya Pradesh, Hindu Thakurs and Brahmans, and Muslim Pathans own substantial land, while lower-ranking Weaver (Koli) and Barber (Khawas) caste members and others own smaller farms. In many areas of the south, Brahmans are major landowners, along with some other relatively high-ranking castes. Generally, land, prosperity, and power go together.

In some regions, landowners refrain from using plows themselves but hire tenant farmers and laborers to do this work. In other regions, landowners till the soil with the aid of laborers, usually resident in the same village. Fellow villagers typically include representatives of various service and artisan castes to supply the needs of the villagers—priests, carpenters, blacksmiths, barbers, weavers, potters, oilpressers, leatherworkers, sweepers, waterbearers, toddy-tappers, and so on. Artisanry in pottery, wood, cloth, metal, and leather, although diminishing, continues in many contemporary Indian villages as it did in centuries past. Village religious observances and weddings are occasions for members of various castes to provide customary ritual goods and services in order for the events to proceed according to proper tradition.

Aside from caste-associated occupations, villages often include people who practice nontraditional occupations. For example, Brahmans or Thakurs may be shopkeepers, teachers, truckers, or clerks, in addition to their caste-associated occupations of priest and farmer. In villages near urban areas, an increasing number of people commute to the cities to take up jobs, and many migrate.

Some migrants leave their families in the village and go to the cities to work for months at a time. Many people from Kerala, as well as other regions, have temporarily migrated to the Persian Gulf states for employment and send remittances back to their village families, to which they will eventually return.

At slack seasons, village life can appear to be sleepy, but usually villages are humming with activity. The work ethic is strong, with little time out for relaxation, except for numerous divinely sanctioned festivals and rite-of-passage celebrations. Residents are quick to judge each other, and improper work or social habits receive strong criticism. Villagers feel a sense of village pride and honor, and the reputation of a village depends upon the behaviour of all of its residents.

Village Unity and Divisiveness

Villagers manifest a deep loyalty to their village, identifying themselves to strangers as residents of a particular village, harking back to family residence in the village that typically extends into the distant past. A family rooted in a particular village does not easily move to another, and even people who have lived in a city for a generation or two refer to their ancestral village as "our village."

Villagers share use of common village facilities—the village pond (known in India as a tank), grazing grounds, temples and shrines, cremation grounds, schools, sitting spaces under large shade trees, wells, and wastelands. Perhaps equally important, fellow villagers share knowledge of their common origin in a locale and of each other's secrets, often going back generations. Interdependence in rural life provides a sense of unity among residents of a village. A great many observances emphasize village unity. Typically, each village recognizes a deity deemed the village protector or protectress, and villagers unite in regular worship of this deity, considered essential to village prosperity.

They may cooperate in constructing temples and shrines important to the village as a whole. Hindu festivals such as Holi, Dipavali (Diwali), and Durga Puja bring villagers together. In the north, even Muslims may join in the friendly splashing of colored water on fellow villagers in Spring Holi revelries, which involve villagewide singing, dancing, and joking. People of all castes within a village address each other by kinship terms, reflecting the fictive kinship relationships recognized within each settlement. In the north, where village exogamy is important, the concept of a village as a significant unit is clear. When the all-male groom's party arrives from another village, residents of the bride's village in North India treat the visitors with the appropriate behaviour due to them as bride-takers—men greet them with ostentatious respect, while women cover their faces and sing bawdy songs at them.

A woman born in a village is known as a daughter of the village while an in-married bride is considered a daughter-in-law of the village. In her conjugal home in North India, a bride is often known by the name of her natal village; for example, Sanchiwali (woman from Sanchi). A man who chooses to live in his wife's natal village—usually for reasons of land inheritance—is known by the name of his birth village, such as Sankheriwala (man from Sankheri).

Traditionally, villages often recognized a headman and listened with respect to the decisions of the *panchayat*, composed of important men from the village's major castes, who had the power to levy fines and exclude transgressors from village social life. Disputes were decided within the village precincts as much as possible, with infrequent recourse to the police or court system. In present-day India, the government supports an elective *panchayat* and headman system, which is distinct from the traditional council and headman, and, in many instances, even includes women and very low-caste members. As older systems of authority are challenged, villagers are less reluctant to take disputes to court.

The solidarity of a village is always riven by conflicts, rivalries, and factionalism. Living together in intensely close relationships over generations, struggling to wrest a livelihood from the same limited area of land and water sources, closely watching some grow fat and powerful while others remain weak and dependent, fellow villagers are prone to disputes, strategic contests, and even violence. Most villages include what villagers call "big fish," prosperous, powerful people, fed and serviced through the labors of the struggling "little fish." Villagers commonly view gains as possible only at the expense of neighbors. Further, the increased involvement of villagers with the wider economic and political world outside the village via travel, work, education, and television; expanding government influence in rural areas; and increased pressure on land and resources as village populations grow seem to have resulted in increased factionalism and competitiveness in many parts of rural India.

Social Inequality and Exclusion

In every society some people have a greater share of valued resources-money, property, education, health and power than others. These social resources can be divided into three forms of capital-economic capital in the form of material assets and income; cultural capital such as educational qualifications and status; and social capital in the form of networks of contacts and social associations. Often these three forms of capital overlap and one can be converted into the other. For example a person from a well-off family can afford expensive higher education and so can acquire cultural or educational capital. Patterns of unequal access to social resources are commonly called social inequality. Social inequality reflects innate differences between individuals for example their varying abilities and efforts. Someone may be endowed with exceptional intelligence or talent or may have worked very hard to achieve their wealth and status. However by and large social inequality is not the outcome of innate or natural differences between people but is produced by the society in which they live.

Sociological Importance of Village in India

"India lives in its villages"-Mahatma Gandhi.

Literally and from the social, economic and political perspectives the statement is valid even today. Around 65% of the State's population is living in rural areas. People in rural areas should have the same quality of life as is enjoyed by people living in sub urban and urban areas. Further there are cascading effects of poverty, unemployment, poor and inadequate infrastructure in rural areas on urban centres causing slums and consequential social and economic tensions manifesting in economic deprivation and urban poverty. Hence Rural

Development which is concerned with economic growth and social justice, improvement in the living standard of the rural people by providing adequate and quality social services and minimum basic needs becomes essential.

The present strategy of rural development mainly focuses on poverty alleviation, better livelihood opportunities, provision of basic amenities and infrastructure facilities through innovative programmes of wage and self-employment. The above goals will be achieved by various programme support being implemented creating partnership with communities, non-governmental organizations, community based organizations, institutions, PRIs and industrial establishments, while the Department of Rural Development will provide logistic support both on technical and administrative side for programme implementation. Other aspects that will ultimately lead to transformation of rural life are also being emphasized simultaneously.

Though the percentage of persons below poverty level in Tamil Nadu has come down significantly between 1993-94 (35.03%) and 1999-2000 (21.12%) as a result of the implementation of various Central and State sponsored schemes, the level of poverty both in absolute numbers (130.40 lakh persons) and percentage of population below poverty line (21.12%) in Tamil Nadu is highest among the four southern States. In spite of huge investments on wage and self-employment programmes, the level of unemployment as per the NSSO 55th round (1999-2000) for Tamil Nadu compared to All India is the second highest among major States in 1987-88 and 1993-94 and third highest in 1999-2000.

The Government's policy and programmes have laid emphasis on poverty alleviation, generation of employment and income opportunities and provision of infrastructure and basic facilities to meet the needs of rural poor. For realising these objectives, self-employment and wage employment programmes continued to pervade in one form or other. As a measure to strengthen the grass root level democracy, the Government is constantly endeavouring to empower Panchayat Raj Institutions in terms of functions, powers and finance. Grama sabha, NGOs, Self-Help Groups and PRIs have been accorded adequate role to make participatory democracy meaningful and effective.

In India the importance of rural sociology gained recognition after independence. The agrarian context occupies special status both in the social scientific literature on India and in the literature on agrarian societies in general. However unlike studies on caste, kinship, village community, gender, study of agrarian relations did not occupy a central position in Indian sociology. The first systematic study of rural India was done by D.N Majumdar followed by N.K Bose, S.C Dubey, M.N Shrinivas. However it was with the publication of Andre'Be'teille's Studies in Agrarian Social Structure in 1974 that agrarian sociology gained professional respectability within the two disciplines.

Peasant studies in a way arrived in India with village studies. The collection of essays, Village India, edited by Marriot with its emphasis on little communities and great communities was brought out under the direct supervision of Robert Redfield. By defining little communities not in relation to land but through other social institutions such as kinship, religion and the social organization of caste there was a shift away from looking at the rural population in relation to agriculture and land. Caste hierarchy came to be defined in terms of ritual or social interaction over institutions of commensality and marriage.

According to Nelson up to the comparatively recent times the story of man is largely the story of rural man. So rural society is the basic foundation of human life, the keystone of the developmental process and the basic unit of social structure. Villages have been in existence since time immemorial unlike cities which are of more recent origin. In the Indian context rural sociology is of greater significance of the following reasons.

According to S.C Dubey from time immemorial village has been a basic and important unit in the organization of Indian social life. Unique nature of transformation of Indian society where elements of traditional and modern cultures have been juxtaposed. For rural development and solution of rural problems according to A.R Desai this systematic study of rural organization of its structure; function and evolution has not only become necessary but also urgent after the advent of independence. Growing influence of industrialization and urbanization. Village as the basic unit of study. Scientific study of village community is a prerequisite for democratic decentralization.

In modern India, the need of rural sociology is very urgent and it is progressive social science gaining importance.

Rural Urban Contrast

Many families and individuals find themselves, at least at some point, questioning the advantages of rural versus urban life. Quality of life is one of the central issues to consider in any comparison between rural versus urban living. While a case can be made for either location as being the best place to live, it is worthwhile to consider how these two options, rural versus urban, are similar and different. Important factors such as the capacity to make general choices, diversity, health, and employment concerns all influence both sides of the comparison and although each both rural and urban living offer great benefits, they both have a seemingly equal number of drawbacks. Rural and urban areas are generally similar in terms of terms of human interaction but differ most widely when diversity and choice are issues.

There are a number of positive as well as negative factors that contribute the overall quality of life in urban centres and if there is any general statement to be made about urban living, it is that there is a great deal of diversity and choice. In urban areas, there are many more choices people can make about a number of aspects of their daily lives. For instance, in urban areas, one is more likely to be able to find many different types of food and this could lead to overall greater health since there could be a greater diversity in diet. In addition, those in urban areas enjoy the opportunity to take in any number of cultural or social events as they have a large list to choose from. As a result they have the opportunity to be more cultured and are more likely to encounter those from other class, cultural, and ethnic groups.

Parents have a number of choices available for the education of their children and can often select from a long list of both public and private school districts, which leads to the potential for better education. It is also worth noting that urban areas offer residents the possibility to choose from a range of employment options at any number of companies or organizations. Aside form this, urbanites have better access to choices in healthcare as well and if they suffer from diseases they have a number of specialists to choose from in their area. According to one study conducted in Canada, "rural populations show poorer health than their urban counterparts, both in terms of general health indicators (i.e. standardized mortality, life expectancy at birth, infant mortality) and in terms of factors such as motor vehicle accidents and being overweight" (Pampalon 421). This could be the result of less reliance on vehicles in urban areas as well as greater emphasis on walking. Despite the conclusions from this study, however, there are a number of drawbacks to urban living as well. Although the life expectancy in cities may be higher, pollution (noise and atmospheric) is an issue that could impact the overall quality of life. In addition to this, overpopulation concerns can also contribute to a decrease in the standard of living.

Rural places do not offer the same level of choice and in very isolated areas and one might be forced to commute long distances to find even a remote selection of the diversity found in urban centres. Still, despite this lack of choice, there are a number of positive sides to rural living in terms of quality of life.

For instance, living in a rural area allows residents to enjoy the natural world more easily instead of having to go to parks. In addition, people do not have to fight with the daily stresses of urban life such as being stuck in traffic, dealing with higher rates of crime, and in many cases, paying higher taxes.

These absences of stressors can have a great effect on the overall quality of life and as one researcher notes, "People living in rural and sparsely populated areas are less likely to have mental health problems than those living in urban areas and may also be less likely to relapse into depression or mental illness once they have recovered from these in more densely populated areas".

The lack of daily stress found in cities from external factors (traffic, long lines, feeling caged, etc) has much to do with this. While there may not be a large number of stores and restaurants to choose from, those in rural areas have the benefit of land upon which to grow their own food, which is much healthier. Although urban populations have large numbers of social networks and networking opportunities, rural communities offer residents the ability to have long-lasting and more personal relationships since they encounter the same people more frequently. While there are not as many schools to choose from and sometimes rural schools are not funded as well as some others, children can grow up knowing their classmates and experience the benefits of smaller classrooms.

One of the drawbacks to living in a rural area, however, is that unlike urban areas, residents do not have the best opportunity to choose from a range of employment options. While they can commute to larger towns, this gets expensive and is not as convenient as working close to their residence. In general, if there is any statement to be made about the quality of life of rural living, it is that there is a greater ability to connect with people and the landscape. The quality of life in urban areas is similar to that in rural areas in that both involve a high degree of socialization, even if on a cursory level. Where they differ most noticeably is in the availability of choices and diversity, especially when vital factors (healthcare, education, and employment options) are concerned.

Change in Indian Rural Life

In many ways independence from colonial rule in 1947 marked the beginning of a new phase in the history of Indian agriculture. Having evolved out of a long struggle against colonial rule with the participation of the people from various social categories, the Indian state also took over the task of supervising the transformation of its stagnant and backward economy to make sure that the benefits of economic growth were not monopolized entirely by a particular section of society. It is with this background that development emerged as a strategy of economic change and an ideology of the new regime.

However at the micro-level the structures that evolved during colonial rule still continued to exist. The local interests that had emerged over a long period of time continued to be powerful in the Indian countryside even after the political climate had changed. According to Daniel Thorner the earlier structure of land relations and debt dependencies where a small section consisting of few landlords and money lenders were dominant continued to prevail in the Indian countryside. The nature of property relations, the local values that related social prestige negatively to physical labour and the absence of any surplus with the actual cultivator for investment on land ultimately perpetuated stagnation. This complex of legal, economic and social relations typical of Indian countryside served to produce an effect that Thorner described as a built-in depressor.

Rural Religion

Many Indian villages have Brahmanic temples within them, however the religious focus is mainly on the shrines of the village's goddess and god. Rural Indians inhabit a world full of divine and semi-divine beings; tree spirits (yakshas), ghosts (bhootas), puranic, local, personal and ancestral gods who co-exist in a complex hierarchy. As well as public shrines to the gods, every Hindu home has a domestic shrine and a veneration of sacred trees and snakes were both attested to millennia ago and still play a important part in religious practice and belief today.

Unlike in orthodox puranic Hinduism villagers have direct access to the local gods and do not require the intercession of a priest. The Goddess also plays a larger role in local religion and rural religion is centred on specific places of perceived spiritual power. The shrines themselves are relatively simple affairs. They are usually covered and often enclosed on three sides by a low wall. Shrines with buildings on them are quite rare. The simplest form of shrine can consist of a pile of stones by a riverbank or in a field, where, at some point in the past a spirit had made its presence known. Villages often have a number of shrines to different deities located at the edge of the village.

In South India the most visually striking shrines are those dedicated to the god Aiyanar. Aiyanar is a Brahmanised pre-Aryan deity and is regarded as a benign being concerned with the welfare of the village community rather than that of the individual. Shrines to Aiyanar are used on special occasions and can therefore sometimes look neglected. Aiyanars power is dependent on three subordinate gods named Karrupu, Muniyan and Maturaiviran who are regarded as impure spirits.

Aiyanars shrines sometimes contain huge brightly painted terracotta statues of these subordinate gods. Throughout India, a part of ritual practice involves making offerings of food, flowers, incense or terracotta figures, mainly of animals, to the deities. Life size terracotta horses are offered to Aiyanar and are believed to serve the god in the spirit world and carry him around the village at night as he protects it. The horses are renewed from time to time but the old ones are kept so the shrines often contain a small herd. Devotional terracottas are made by professional potters of the Kumhars caste and have a long history going back to the Harrapan culture of 2,500 BCE. The skill of a potter is believed to be a gift of the Gods and the potters are understood to possess a kind of magical power.

Another widely worshipped God in South India is Murukan (the Tamil name of Kartikeya). Murukan is invoked for protection and is a god of war associated with the planet Mars. In village shrines, Murukan is represented by a spear. In areas where he is regarded as the offspring of Shiva and Parvati, a trisul (or trident, an emblem of Shiva) can be found next to the spear. Offerings of ghee or fruit are skewered on to the blade tips although human hair is considered to be the most auspicious offering to Murukan. In addition to the type of offerings made to the male gods the village goddesses can require placating with live offerings. Where Goddess shrines contain anthropomorphic images of the deity, they are usually roughly carved in stone or wood, garlanded with flowers and sometimes dressed in clothing.

An unpaid priest and his assistants have the duty to maintain the shrines (at the community's expense) and to propitiate the deity to ward off communal bad luck and disease. Individual villagers, regardless of caste, can approach the village deity directly as and when they have a need. At specific times of year and during crisis a festival is held in honour of the Goddess. The main feature of these festivals is the sacrifice of an animal, at one time buffalo sacrifice was widely practised, nowadays the victim is more likely to be a goat or chicken. In Northern India the blood from such sacrifices is shunned because it is regarded as impure.

In the South the opposite is true, the blood is liberally flecked and smeared over the altar and devotees, as there is a strong belief in the power of the impure. The meaning of the sacrifice appears to be to release the power of the Goddess into the community. As the Goddess is regarded as both destructive and creative it is not considered wise to allow her presence in the village for very long. Spirit possession is also a feature of the Goddess festival as are severe penances such as hook hanging, fire walking and fire swallowing. Brahmins discourage the practice of animal sacrifice and not all villages Goddesses demand it. Examples of 'vegetarian' Goddesses in Tamil Nadu, include Antal and Sapta Kannimar, while in Orissa Ma Ksetrapala prefers cannabis. Widely known Goddesses such as Mariyamman in Tamil Nadu and Sitala in North India are relived differently from one area or village to the next and offering vary according to their perceived character. In parts of South India Goddess shrines are located to the north of the village. This is significant as the north is associated with spiritual knowledge and disease and so emphasises the innate duality of the Goddess. Western scholars have tended to portray village Goddesses as sinister or malicious and venerate or placated through fear, however they are best described as ambivalent and are the objects of intense devotion.

Snake veneration is also an important part of Indian religion. An important festival called the Nagapanchami is held at the start of the rainy season in August and it is a common sight to see Naga worship all over India by those hoping to be blessed with children. The Nagas are a race of serpents whose origins are described in the Mahabharata and the Varaka Purana. As divine beings they are depicted in human, half human and fully serpent form. In rural areas snakes are commonly found near ant hills and termite mounds, both of which are regarded as the entrances to the otherworld. The mounds are frequently marked with ash and offerings of milk or eggs are made to the resident Naga. Particularly impressive mounds can thatched roofs supported by posts placed over the, similar to some Goddess shrines. Like the dragons of European mythology Nagas are seen as guardians of the otherworld and of treasure. There is also an association between snakes and trees, both being symbols of fertility and the roots of trees, like the termite mound, is seen as an entrance to the otherworld. In South India women desiring children erect snake stones under sacred tress.

These stones have stylised cobras carved on them represent the Goddess Nakamal (snake virgin) and are immersed in water for several months to empower them then erected under Nim, or Pipal trees, accompanied by prayer and ritual.

In North India Manasa is invoked for protection against snakebite and to cure infertility and is represented by an earthen snake image, the branch of a tree or a water pot. As mentioned above, trees are considered to be symbols of fertility however overlying this aspect of folk religion is the Vedic concept of the tree as the axis of the Cosmos. The Rg Veda relates how sacrificial offerings were tied to a tree so that the energies released would travel upwards to the realm of Gods. Sacred trees can be found in villages, within a temple complex or shrine and in forests. The tree spirits or Yaksha (male) Yakshini (female) was worshipped in very early times, however much of the reverence shown to them has been transferred to the river Goddesses. The Yaksha is still recognised but villagers can make offerings to their ancestors, local Gods and puranic deities through the tree. The Pipal tree, for example, is believed to be the home of the elder sister of Laksmi, a Goddess named Nairrti, who seems to be a Brahmanised village Goddess. Nairrti could be described as Laksmi's dark side, as she represents misfortune. The Pipal is not touched except on Saturday when Nairrti is believed to visit Laksmi. Shrines to Nairrti are found outside villages. The Gond tribe venerates pairs of trees simply as "man" and "woman". The shade of a tree is also considered sacred. The ancient art of mediumship under sacred trees, especially those in the grounds of temples and shrines is still a common occurrence. In forests trees can be found stained and garlanded with beads, representing Bana Durga (Forest Durga) and shrines to forest Kali can be found under pairs of trees. The trees are given a similar reverence and type of offerings as those given to the Gods.

Sacred spaces are created in the home, often an entire room is used as a place of worship. Offerings of food and incense are made to the Gods, who are represented by the statues made from brass, wood or stone. The help of the deity is sought mainly but not exclusively by the women of the house for a range of personal and domestic issues. The custom of making a sacred vow or Vrat to the deity forms the basis for many domestic rituals. The custom of Vrat originates in part from the Artharvaveda, the last of the four great Vedas which contains spells and incantations derived from folk religion. The knowledge of Vrat rituals is said to have been spread throughout India by magician priests who were accompanied by a sacred prostitute and a musician as they toured villages in brightly painted carts. The success of the ritual relies upon an implicit belief in magic and the focusing of intent, empowered by gesture and incantation to raise energy. The nature of the rituals varies according to the individual, although they do usually involve recitation of mantras and the making of magical drawings. The ritual may also need repeating on a monthly or seasonal basis. Finally a vow is made to the deity to perform an austerity, usually fasting for a fixed period, for example every Monday for sixteen weeks, or to provide gifts when the outcome is known.

Rural religion and in particular the local Goddess has tended to be ignored or denigrated by both Indian and Western scholars and regarded as a subject not worthy of serious scholarly investigation. This brief article hopes to have shown that rural religion and the day to day worship of local Gods and Goddesses forms the basis of religious activity in India and provides Hinduism with much of its continuity with India's ancient past.......

Rural Festivals (Cultural Aspect)

Indian Villages celebrate some of the unique festivals that reflect the rural charm and simplicity of the Indian people. The villages of the Indian states are special for their distinguished fairs and festivals, however, festivals like Republic Day, Diwali, Gandhi Jayanti, Id-ul-Fitr, Independence Day and Janmastami are celebrated nationwide. Besides the religious festivals cultural ones are also predominant in the Indian villages. The Indian Village festivals according to the location of the villages are as follows:

North India Village Festivals-North India comprises the villages of Delhi, Jammu and Kashmir, Himachal Pradesh, Punjab, Haryana, Uttarakhand and Uttar Pradesh. The composite culture and the festivals of North India are closely associated with the Himalayas and sacred rivers, passing across the states. Most of the festivals celebrated in these villages are common and similar in their themes. Karva Chauth, Vasant Panchami, Diwali, Lohri, Buddha Purnima, Kheer Bhawani are the commonly celebrated all across northern India.

East India Village Festivals-East Indian states of West Bengal, Bihar, Jharkhand and Orissa comprise the villages in this region. Cuisine plays a vital role in the eastern Indian festivals. An important feature of the festivals here is that these are diverse. While the most popular festivals celebrated in the villages of West Bengal are the Durga Puja and Kali Puja, Ratha Yatra is celebrated with lot of fervour in Orissa. The typical rural festivals of eastern India are Jatra

Festival, Jhoolan, Poush Mela and Vasanta Utsav. Cultural festivals are also an important part of the East Indian village festivals.

North-East India Village Festivals-The northeastern states of India are Sikkim, Nagaland, Meghalaya, Mizoram, Arunachal Pradesh, Assam, Tripura and Manipur. The culture of these northeastern villages vastly depends on the migrated tribal customs and traditions. The villages of Mizoram, Meghalaya and Nagaland celebrate some tribal festivals like Chapchar Kut, Mim Kut, Ningol Chakouba, Heikru Hitongba among many others.

South India Village Festivals-The villages of South India belong to the states of Kerala, Tamil Nadu, Karnataka, Andhra Pradesh, Goa and Maharashtra. The South Indian culture mostly includes festivals that are related to their coconut preparations, religion and water games; their common festivals are Onam, Pongal and numerous festivals on music and dance are quite popular in south Indian villages. The Andaman and Nicobar Islands are into several tribal festivals, due to their major tribal population.

Central India Village Festivals-The Central Indian villages belong to the states of Madhya Pradesh and Chhattisgarh. Arwa Teej, Kajri Navami, Bhojali and Chherta are the common festival of the rural areas in central India. Splendor, traditional songs, dances and colourful dresses are indispensable from these Indian village festivals.

West India Village Festivals-The West Indian states of Rajasthan and Gujarat have some of the most colorful and cultural villages, celebrating the traditional festivals. These festivities date back to the customs of the early raja and maharaja eras. Besides celebrating the popular Hindu festivals, Jain and Buddhist festivals are also integrated in the culture of these villages.

India is a land of unique festivals, retaining its culture and historical significance; the Indian villages are no exception. The rural Indian boasts some of the oldest and exceptional traditions that have grown as distinguished festivals that not only serve entertainment, but also speaks volumes about the Indian heritage and history. The geographic divisions of India definitely divide the language, rituals and festivals. However, the spirit with which the Indian village festivals are celebrated remain, predominantly, similar.

Rural Family

Indian family structure is believed to be the unit that teaches the values and worth of an honest living that have been carried down across generations. Since the Puranic ages, Indian family structure was that of a joint family, indicating every person of the same clan living together. However, this idea of elaborate living disintegrated in smaller family units.

In India, people learn the essential themes of cultural life within the bondings of a family. In ancient days, the basic units of society had been the patrilineal family unit with wider kinship groupings. The most widely preferred residential unit is the joint family, ideally consisting of three or four patrilineally related generations, all living under one roof, working, worshiping, eating, and cooperating together in communally beneficial social and economic activities. Patrilineal joint families include men related through the male line, along with their wives and children.

The young married women live with their husband‘s relatives after marriage, but they retain important bonds with their natal families as well. Despite the continuous and growing impact of urbanization, secularization, and Westernization, the conventional joint household of Indian family structure, both in ideal and in practice, remains the chief social force in the lives of Indians. Loyalty to family is a deeply imbibed in every member of the family. Large families eventually faced difficulties to suit with modern Indian life. The modern style of livinf, modern occupations and beliefs are eventually confronting problems to get adjusted. The joint family is now quite unfamiliar in cities. However, the relative ties are maintained within the kinships, since these very ties can prove to be crucial while any kind of emergency. Numerous prominent Indian families, such as the Tatas, Birlas, and Sarabhais, retain joint family arrangements even today and they work together to control some of the country‘s largest financial empires.

The Indian joint family structure is an ancient phenomenon, but it has undergone some change in the late twentieth century. Living arrangements vary widely depending on region, social status, and economic circumstance. With the passing time, nuclear families have evolved that is a couple living with their unmarried children. There are often strong networks of kinship ties through which economic assistance and other benefits are obtained. Often clusters of relatives live near each other, who are easily available and respond to the give and take of kinship obligations. Even when relatives cannot actually live in close proximity, they typically maintain strong bonds of kinship and attempt to provide each other with economic help, emotional support, and other required benefits.

The Indian joint families grew even larger and finally they divide into smaller units, passing through a expected cycle over time. The breakup of a joint family into smaller units does not necessarily symbolize the rejection of the joint family ideal. Rather, it is usually a reaction to a variety of conditions, including the requirement for some members to move from village to city, or from one city to another to obtain the advantage of employment opportunities.

Splitting of the family is often blamed on quarrelling women, the wives of co-resident brothers and so on. Although women's disputes may, in fact, lead to family division, men's disagreements are responsible as well. Despite cultural ideals of brotherly harmony, adult brothers often quarrel over land and other matters, leading them to decide to live under separate roofs and split their property. Frequently, a large joint family divides after the death of elderly parents, when there is no longer a solitary authority figure to hold the family factions together. After division, each new housing unit, in its turn, usually comes together when sons of the family marry and bring their wives to live in the family home. Some Indian family structure bears special mention because of their unique qualities. In the sub-Himalayan region of Uttar Pradesh, polygyny is generally practiced. A polygynous family comprises a man, his two wives, and their unmarried children. Various other Indian family structures occur there, including the supplemented subpolygynous household, where a woman whose husband lives elsewhere, stays with her children and other adult relatives. Among the Buddhist people of the mountainous Ladakh District of Jammu and Kashmir, fraternal polyandry is practiced; a household may include a set of brothers with their common wife or wives. This family type, in which brothers also share land, is almost certainly linked to the extreme scarcity of cultivable land in the Himalayan region, because it discourages disintegration of holdings.

The inhabitants of the northeastern hill areas are known for their matriliny order that distinguish the decent and inheritance of a family in the female line rather than the male line. One of the largest of these groups, the Khasis of Meghalaya is divided into matrilineal clans. Here, the youngest daughter receives almost all of the inheritance including the house. A Khasi husband lives in his wife's house.

Perhaps the strangest Indian family structure form is the traditional Nayar taravad, or great house. The Nayars are a cluster of castes in Kerala who are high-ranking and prosperous. The Nayars maintained matrilineal households in which sisters and brothers and their children remain as the permanent residents. After an official childhood marriage, each woman received a series of visiting husbands in the taravad. Her children were all considered as the legitimate members of the taravad.

The eldest brother of the senior woman managed property, matrilineally inherited. This kind of Indian family structure has been eleminated in the twentieth century, and in the 1990s probably fewer than 5 percent of the Nayars still live in matrilineal taravads. Like the Khasis, Nayar women are well educated and powerful within the family. Malabar rite Christians, an ancient community in Kerala, adopted many Indian family structure practices alike their powerful Nayar neighbors, including naming their sons for matrilineal descent. Their relationship system, however, is patrilineal. Thus, Indian family structure has been varied in varied periods of time and in different regions of the nation. The society structure and regulations have the highest influence on such formations of Indian family structures.

Joint Family System

A Hindu Joint Family or Hindu undivided family (HUF) or a Joint Family is an extended family arrangement prevalent among Hindus of the Indian subcontinent, consisting of many generations living under the same roof. All the male members are blood relatives and all the women are either mothers, wives, unmarried daughters, or widowed relatives, all bound by the common sapinda relationship.

The joint family status being the result of birth, possession of joint cord that knits the members of the family together is not property but the relationship. The family is headed by a patriarch, usually the oldest male, who makes decisions on economic and social matters on behalf of the entire family. The patriarch's wife generally exerts control over the kitchen, child rearing and minor religious practices. All money goes to the common pool and all property is held jointly.

There are several schools of Hindu Law, such Mitakshara, the Dayabhaga, the Murumakkattayam, the Aliyasanthana etc. Broadly, Mitakshara and Dayabhaga systems of laws are very common. Family ties are given more importance than marital ties. The arrangement provides a kind of social security in a familial atmosphere. Due to the development of Indian Legal System, of late, the female members are also given the right of share to the property in the HUF. In CIT vs Veerappa Chettiar, 76 ITR 467 (SC), Supreme Court had occasion to decide on an issue whether after the death of all the female members in a HUF, the HUF would still exist.

Six key aspects of Joint Family are:-

- head of the family takes all decision
- all members live under one roof
- share the same kitchen
- three generations living together (though often two or more brothers live together, or father and son live together or all the descendants of male live together)
- income and expenditure in a common pool-property held together.
- a common place of worship
- all decisions are made by the male head of the family-patrilineal, patriarchal.

Indians identify themselves with a particular religion but also affiliate themselves with a specific geographical region or state in India. Religion specifies the form of worship and guides their dayto-day behaviour, while the specific region generally identifies the language one speaks, the literature, art, music one prefers, the food one eats, and the clothing one wears (Segal 1991).

Because India is a secular and ethnically diverse society, there are religious, regional, cultural, social, and educational variations in structural and functional patterns of family life. Hence, it is difficult to generalize values, behaviours, attitudes, norms, mores, practices, traditions, and beliefs about family life from one community to all Indian communities. Because the large majority of Indians are Hindus, this chapter will primarily focus attention on family life in the Hindu community.

The Hindus believe in a multitude of gods and goddess that are an integral aspect of Hindu mythology. Hinduism is a major world religion, has approximately 800 million followers, and also has had a profound influence on many other religions during its long history that dates back to 1500 B.C.E. The ideal Hindu lifestyle is influenced by the teachings in the Upanishads, Vedas, Bhavadgita, Ramayana, and Mahabharata. These scriptures stress the importance of work, knowledge, sacrifice, and service to others and finally, the renunciation of worldly goods in later life (Chekki 1996). Hinduism is not an organized religion like Western religions (Nandan and Eames 1980), but rather a way of life. According to the Hindu ideology, a person's life consists of four stages that correspond with the human life-cycle stages. The first stage is the Brahamacharya ashram (apprenticeship)—

this is the period of discipline and education. The second stage is the Grihastha ashram (household and family), devoted to marriage, parenthood, family, and establishment of a household. Stage three is the Vanaprastha ashram (gradual retreat) and is characterized by a gradual retreat and loosening of social, emotional, and material bonds. Finally, the goal of the fourth and final stage, the Sanyasa ashram (renouncement), is to seek solitude, indulge in meditation, prepare for death, and strive for salvation and wisdom. Most Hindu households have a prayer platform or room that is considered the most sacred place in the home. Most devout Hindus are vegetarians. They pray, fast, and worship their deity at least once a day, especially on holy days and days of festivities. As part of the religious activities, Hindus take regular morning baths, recite and chant certain mantras, light incense, prepare specific food items, offer flowers to the deities, and worship ancestors.

Arranged Marriages in India

People of India basically follow the arranged marriage system, and they consider it as something great. Dating is a taboo in that country. However, it has its own merits and demerits. Indian people give much importance to family relationship. The system seems to protect the family. The parents take care of their children, and the children obey their parents. Parents find suitable spouses for their children from appropriate families. So, there is no chance of marrying outside their own religion, caste, social status or economic class. This protects the couple from the problems that usually originate from disparity of religion, caste and class. Through a marriage two families come into mutual relationship, and both families together try to work out the marriage if problems arise in the marriage.

Nevertheless, the arranged marriage system has its flaws. This system originated when child marriage was the custom in India. Children at an early age, even before their puberty, were given in marriage. Such children could not give valid consent to marriage, and so parents were consenting. The purpose of child marriage was to prevent those children from seeking by themselves (when they become adults) somebody from lower caste or lower class for marriage. It was a means of restriction to their children from marrying outside their race and social status. Thus arranged marriage system is a product of caste system. It has developed to promote racism and classism, and it is not based on any spiritual value.

Child marriages are now abolished by law, and the children are free to choose their own partners, according to law. But, you know, racism is in the blood, and the parents, even now, try to control their children by arranging marriages within the limits of race, caste, class and religion. If children find their own mates, parents would threaten them in many ways — threatening not to give them any share of family property or wealth; threatening to drive them out of their own homes. If any children marry according to their own desire, parents would consider it as a threat and shame to the family. So, many men and women just accept what their parents arrange for them. They don't want to lose their share of property, and they don't want to invite any shame to their family. Even if they don't like the spouse they get, they accept what they receive and suffer the consequences silently. According to divine plan marriage should happen through love and the consent of those who marry. In arranged marriages, it is the parents who decide and give consent. Very often there is so much force and fear involved in marriage—force from the parents and fear from the part of children who wish to marry. It doesn't fit into the modern definition of marriage which is the total partnership of the whole of life which happens through mutual consent and love of those who enter into marital union. Marriage should happen through mature decision of those who marry, and not of their parents.

It also should be noted that arranged marriages are prevalent among the high-caste and high-class people. They are the people who want to protect their "status". People of lower strata do not care about this very much. The reason for this is: they have nothing to lose. However, they also try to imitate the way of higher level people, believing that it is something great. Another reason for arranged marriages among the lower class is ethnic rivalry and pride over their own race. Matrimonial classifieds in newspapers or help of marriage brokers are sought in arranged marriage when the family fails to find "suitable" spouses for their children. All necessary "qualifications" (racial, religious, economic, educational, etc.) of the "candidates" would be stated in the advertisement.

A modern curse connected with arranged marriage is dowry. It is a social evil in India though it is prohibited by law. When they arrange a marriage, the consent of boy and girl who are to enter into marriage is not important; the negotiation is on the amount of dowry which is to be given by the girl's family. It has become something like a trade in modern Indian culture. The girls from poor families, and the girls who are orphans are not good commodities in this trade; so they remain unmarried. If the promised dowry is not given by due date, the girl would be persecuted and thrown out by her husband's family; or, she would die in a "domestic accident". Do not think that I am exaggerating; it is happening in India everyday.

Another drawback of arranged marriage is that the partners to marriage do not know their future spouses before marriage. In arranged marriage it is not important at all. In many cases, the boy and girl who get ready to enter into marriage may see each other two or three times before marriage, and that meeting would be in the presence of parents and other family members. Thus, marriage happens without knowing each other. Many people who work in far away places, especially in gulf countries come home for a leave, and marriage is arranged within a week or two. Legally speaking, you do not give valid consent to accept something you do not know. It is consent that brings a marriage; and, if there is no valid consent, there is no marriage according to the law of the Catholic Church. Marriages contracted because of force or fear would be invalid according to Canon Law. If we strictly analyze, many marriages happening in India are invalid.

Good marriages, that are arranged, do occur. Parents who love their children, and who are not vitiated by false family pride, seek the consent of their children when they are given in marriage. "Good" arranged marriages happen when the parents help their children to find their own partners according to their own desires. The dating that we find in America is a good opportunity for boys and girls to know each other and select their future partners with total freedom and true consent. However, this opportunity is misused by many boys and girls and have brought disaster to their own lives. They totally discard desires of their family. They take dating for total freedom from parents and religion, and total freedom for sex. This has brought much misunderstanding about dating, especially in the circle of people of India.

Arranged marriage system in India is bad in one sense but good in another sense. It is bad when marriage is arranged with such a hatred and prejudice over other religions, castes and races; it is bad when parents over-protect and control their children to the extent of denying every wish, and even every right of their children in choosing their partners. Arranged marriages are wonderful when parents and children love each other sincerely, and total freedom is given to children for final consent to marriage; and, when arrangements are made for the would-be-spouses to meet and to know each other.

Dating system in America is bad when children totally disregard every genuine wish of their parents and consider everything as their freedom and fundamental right to the extent of practicing sex before marriage and considering marriage as a mere contract that can be terminated by a divorce decree. They often have no respect for their religious values at all. According to a friend of mine, dating system is to help the youth to learn to divorce and not to marry. You date with some one, then reject the person and accept another one; so you learn to divorce. It is wonderful when there is sincere attempt to seek the partner for an intimate union. When this happens, there is respect for family and religion; and, they seek parent's advice.

Marrying a poor orphan girl is considered to be wonderful. Doesn't look like giving such a person an asylum? Compassion is a wonderful thing, but marrying somebody out of compassion is not good at all. Marriage is not a charitable work. It is mutual sharing of whole life. Each spouse has to feel equality and mutual respect. There is no place in marriage for superiority and inferiority. There is no meaning of one being submissive and the other being aggressive—although some people believe this to be an ideal marriage.

In India many people believe that by marriage a woman enters into a bondage, and in Indian situation this is pretty accurate—woman is not free. In arranged marriage, her consent is not sought; her desires have no importance; and, even if she loves somebody to be her husband, family not only doesn't give any consideration to that but also threatens her in many ways. After marriage, in many cases, she is like a slave. She must be submissive to the abuses of the husband and his family.

In America, on the contrary, too much freedom is given to person, even to the extent of disregarding the family or religion. However, it should be noted that there is much stress on equality. Though this sense of equality is shown, sometimes to the extent of not having any humbleness to serve the other. Whether it be arranged marriages or courtship marriage, people seek perfect husbands and perfect wives. A perfect husband or a perfect wife is a myth. No such person lives or ever lived in this world. We should not expect anyone to be 100 percent perfect. We are all called to be perfect, but we are only on the way to it. We have to accept each other with each one's weaknesses and failings. As there is no perfect wife or perfect husband, there is no perfect marriage either. Success of marriage is in mutual understanding and acceptance, and also in mutual love and respect.

Marriages in Indian Villages

India is a land of diverse culture and ethnicity, wherein the important occasions are celebrated in myriad ways. Throughout the length and breadth of the country, you will see that wedding is given a paramount importance in people's social life. Here, marriage is not only a legal bonding between and man and a woman, it is an auspicious occasion, which brings the families of the two, closer. The ceremony often takes the shape of a festive occasion that is observed with pomp and gaiety, in the Indian subcontinent. This is the reason why wedding is often celebrated lavishly and elaborately. Nonetheless, you will see a contrast in the ways in which the wedding rituals are observed in the rural and the urban areas of the country.

Weddings in Indian Villages

Just like urban areas of India, wedding is an elaborate affair in the rural parts of the country. The rituals conducted before, during and after wedding are conducted with special attention to the nuances of the same. Depending upon their belief, people in villages organize a number of pujas before the wedding, to ensure the smooth conduct of the ceremony. After the marriage, they conduct certain rituals to ensure that the newly wed couple leads a prospered life forever. If you go deep into the rural areas of the country, you would witness huge differences in the same rituals, because people in the villages are very particular in following the customs that are native to their community. Unlike cities, in Indian villages, the role of wedding planners is played by the family members of the bride and the groom.

The expenses of wedding are generally borne by the bride's parents, while the groom's family arranges the reception party. Since the system of joint families is still prevalent in the villages, the arrangements for a wedding are conducted very smoothly. It is a joint effort of the family members, relatives and close friends, when it comes to organizing a wedding. All the people involved in the preparations for the wedding ensure that everything is conducted smoothly, before, during and after the ceremony.

Depending upon the financial status, the grandeur of the wedding ceremony varies. For instance, if the families of the prospective bride and the groom are not financially well settled, they would restrict the wedding to a short and crisp affair, inviting only the relatives and close friends. On the other hand, if it is the wedding of a renowned person of the village, who is financially well settled, then the wedding would be just like a festival for the village, wherein every villager is invited to mark his/her presence and grace the occasion. In general, weddings in Indian villages are as celebrated with pomp and gaiety, which is no less to those witnessed in urban parts of the country.

Till some times back, people in villages of India were strictly against love marriages. They strongly believed that two people should not tie the wedding knot without the consent of their elder members of the family and relatives. Due to this perception, arranged marriage is customary in Indian villages. However, with the passing times, people living even in the interiors of rural India are not so strictly against the concept of love marriage and hence, if persuaded, they would not hesitate to come forward and solemnize the wedding of the two loving souls. Therefore, the perception of wedding in the rural India has changed drastically.

Sociology of Caste System in India

The Indian caste system describes the system of social stratification and social restrictions in India in which social classes are defined by thousands of endogamous hereditary groups, often termed *jâtis* or castes. Within a jâti, there exist exogamous groups known as gotras, the lineage or clan of an individual. In a handful of sub-castes such as Shakadvipi, endogamy within a gotra is permitted and alternative mechanisms of restricting endogamy are used (e.g. banning endogamy within a surname). The Indian caste system involves four castes and outcasted social groups. Although generally identified with Hinduism, the caste system was also observed among followers of other religions in the Indian subcontinent, including some groups of Muslims and Christians. Caste barriers have mostly broken down in large cities, though they persist in rural areas of the country, where 72% of India's population resides. None of the Hindu scriptures endorses caste-based discrimination, and the Indian Constitution has outlawed caste-based discrimination, in keeping with the secular, democratic principles that founded the nation. Nevertheless, the caste system, in various forms, continues to survive in modern India because of a combination of political factors and social perceptions and behaviour.

History

There is no universally accepted theory about the origin of the Indian caste system. The Indian classes are similar to the ancient Iranian classes ("*pistras*"), wherein the priests are Brahmins, the warriors are Kshatriya, the merchants are Vastriya, and the artisans are Huiti.

Varna and Jati

According to the ancient Hindu scriptures, there are four "varnas". The Bhagavad Gita says varnas are decided based on Guna and

Karma. Manusmriti and some other shastras name four varnas: the Brahmins (teachers, scholars and priests), the Kshatriyas (kings and warriors), the Vaishyas (agriculturists and traders), and Shudras (service providers, laborers).

This theoretical system postulated Varna categories as ideals and explained away the reality of thousands of endogamous Jâtis actually prevailing in the country as being the result of historical mixing among the "pure" Varnas – *Varna Sankara*. All those who did not subscribe to the norms of the Hindu society, including foreigners, tribals and nomads, were considered contagious and untouchables. Another group excluded from the main society was called Parjanya or Antyaja. This group of people formerly called "untouchables", the Dalits, was considered either the lowest among the Shudras or outside the Varna system altogether.

Several critics of Hinduism state that the caste system is rooted in the varna system mentioned in the ancient Hindu scriptures. However, many groups, such as ISKCON, consider the modern Indian caste system and the varna system two distinct concepts. Many European administrators from the colonial era incorrectly regarded the Manusmriti as the "law book" of the Hindus, and thus concluded that the caste system is a part of Hinduism, an assertion that is now rejected by most scholars, who state that it is a social practice, not a religious belief. Manusmriti was a work of reference for the Brahmins of north India, especially Bengal, and was largely unknown in southern India.

Although many Hindu scriptures contain passages that can be interpreted to sanction the caste system, they also contain indications that the caste system is not an essential part of Hinduism. The Vedas placed no importance on the caste system, mentioning caste only once (in the Purush Sukta) out of tens of thousands of verses. Most vedic scholars believe even this to be a subsequent and artificial insertion; B. R. Ambedkar concluded after a thorough study that this is a much later interpolation, giving strong evidence to support his conclusion. In the Vedic period, there was no prohibition against anyone, including the Shudras, listening to the Vedas or participating in any religious rite.

In *Early Evidence for Caste in South India*, George L. Hart stated that "the earliest Tamil texts show the existence of what seems definitely to be caste, but which antedates the Brahmins and the Hindu orthodoxy". He believes that the origins of the caste system can be seen in the "belief system that developed with the agricultural civilization", and was later profoundly influenced by "the Brahmins and the Brahmanical religion". These early Tamil texts also outline the concept of equality. Saint Valluvar has stated "pirapokkum ella uyirkkum", which means "all are equal at birth". Likewise, Saint Auvaiyaar has stated that there are only two castes in the world: those who contribute negatively and those who contribute positively. From these statements, it can be inferred that the caste system is a socio-economic class system.

Caste and Social Status

Traditionally, although the political power lay with the Kshatriyas, historians portrayed the Brahmins as custodians and interpreters of Dharma, who enjoyed much prestige and many advantages.

Fa Hien, a Buddhist pilgrim from China, visited India around 400 AD. "Only the lot of the Chandals he found unenviable; outcastes by reason of their degrading work as disposers of dead, they were universally shunned... But no other section of the population were notably disadvantaged, no other caste distinctions attracted comment from the Chinese pilgrim, and no oppressive caste 'system' drew forth his surprised censure.". In this period kings of Sudra and Brahmin origin were as common as those of Kshatriya *varna* and caste system was not wholly prohibitive and repressive.

The castes did not constitute a rigid description of the occupation or the social status of a group. Since British society was divided by class, the British attempted to equate the Indian caste system to their own social class system. They saw caste as an indicator of occupation, social standing, and intellectual ability. Intentionally or unintentionally, the caste system became more rigid during the British Raj, when the British started to enumerate castes during the ten year census and codified the system under their rule.

The Harijans, or the people outside the caste system, had the lowest social status. The Harijans, earlier referred to as *untouchables* by some, worked in what were seen as unhealthy, unpleasant or polluting jobs. In the past, the Harijans suffered from social segregation and restrictions, in addition to extreme poverty. They were not allowed temple worship with others, nor water from the same sources. Persons of higher castes would not interact with them. If somehow a member of a higher caste came into physical or social contact with an untouchable, the member of the higher caste was defiled, and had to bathe thoroughly to purge him or herself of the impurity. Social discrimination developed even among the Harijans; sub-castes among Harijans, such as the *dhobi* and *nai*, would not interact with lower-order Bhangis, who were described as "outcastes even among outcastes".

Sociologists have commented on the historical advantages offered by a rigid social structure as well as its drawbacks. While caste is now seen as anachronistic, in its original form the caste system served as an instrument of order in a society where mutual consent rather than compulsion ruled; where the ritual rights and the economic obligations of members of one caste or sub-caste were strictly circumscribed in relation to those of any other caste or sub-caste; where one was born into one's caste and retained one's station in society for life; where merit was inherited, where equality existed within the caste, but inter-caste relations were unequal and hierarchical. A well-defined system of mutual interdependence through a division of labour created security within a community. In addition, the division of labour on the basis of ethnicity allowed immigrants and foreigners to quickly integrate into their own caste niches. The caste system played an influential role in shaping economic activities, where it functioned much like medieval European guilds, ensuring the division of labour, providing for the training of apprentices and, in some cases, allowing manufacturers to achieve narrow specialisation. For instance, in certain regions, producing each variety of cloth was the speciality of a particular sub-caste. Additionally, some philosophers have argued that the majority of people would be comfortable in stratified endogamous groups, as they were in ancient times.

Reforms

There have been challenges to the caste system from the time of Buddha, Mahavira and Makkhali Gosala. Opposition to the system of varna is regularly asserted in the Yoga Upanisads and is a constant feature of Cîna-âcâra tantrism, a Chinese-derived movement in Asom; both date to the medieval era. The Nâtha system, which was founded by Matsya-indra Nâtha and Go-rakca Nâtha in the same era and spread throughout India, has likewise been consistently opposed to the system of varna. Many Bhakti period saints rejected the caste discriminations and accepted all castes, including untouchables, into their fold. During the British Raj, this sentiment gathered steam, and many Hindu reform movements such as Brahmo Samaj and Arya Samaj renounced caste-based discrimination.

The inclusion of so-called untouchables into the mainstream was argued for by many social reformers. Mahatma Gandhi called them "Harijans" (children of God) although that term is now considered patronizing and the term Dalit (*downtrodden*) is the more commonly used. Gandhi's contribution toward the emancipation of the untouchables is still debated, especially in the commentary of his contemporary Dr. B.R. Ambedkar, an untouchable who frequently saw Gandhi's activities as detrimental to the cause of upliftment of his people.

The practice of untouchability was formally outlawed by the Constitution of India in 1950, and has declined significantly since then, to the point of a society allowing former untouchables to take high political office, like former President K. R. Narayanan, who took office in 1997, and former Chief Justice K. G. Balakrishnan.

British Rule

The fluidity of the caste system was affected by the arrival of the British. Prior to that, the relative ranking of castes differed from one place to another. The castes did not constitute a rigid description of the occupation or the social status of a group. The British attempted to equate the Indian caste system to their own class system, viewing caste as an indicator of occupation, social standing, and intellectual ability. During the initial days of the British East India Company's rule, caste privileges and customs were encouraged, but the British law courts disagreed with the discrimination against the lower castes. However, British policies of divide and rule as well as enumeration of the population into rigid categories during the 10 year census contributed towards the hardening of caste identities. During the period of British rule, India saw the rebellions of several lower castes, mainly tribals that revolted against British rule. These were:

1. Halba rebellion (1774–79)
2. Bhopalpatnam Struggle (1795)
3. Bhil rebellion (1822–1857)
4. Paralkot rebellion (1825)
5. Tarapur rebellion (1842–54)
6. Maria rebellion (1842–63)
7. First Freedom Struggle (1856–57)
8. Bhil rebellion, begun by Tantya Tope in Banswara (1858)
9. Koi revolt (1859)
10. Gond rebellion, begun by Ramji Gond in Adilabad (1860)
11. Muria rebellion (1876)
12. Rani rebellion (1878–82)
13. Bhumkal (1910).

Modern Status of the Caste System

In some rural areas and small towns, the caste system is still very rigid. Caste is also a factor in the politics of India. The Government of India has officially documented castes and sub-castes, primarily to determine those deserving reservation (positive discrimination in education and jobs) through the census. The Indian reservation system, though limited in scope, relies entirely on quotas. The Government lists consist of Scheduled Castes, Scheduled Tribes and Other Backward Classes:

Scheduled Castes (SC)

Scheduled castes generally consist of "Dalit". The present population is 16% of the total population of India (around 160 million). For example, the Delhi state has 49 castes listed as SC.

Scheduled Tribes (ST)

Scheduled tribes generally consist of tribal groups. The present population is 7% of the total population of India i.e. around 70 million.

Other Backward Classes (OBC)

The Mandal Commission covered more than 3000 castes under OBC Category and stated that OBCs form around 52% of the Indian population. However, the National Sample Survey puts the figure at 32%. There is substantial debate over the exact number of OBCs in India; it is generally estimated to be sizable, but many believe that it is lower than the figures quoted by either the Mandal Commission or the National Sample Survey.

The caste-based reservations in India have led to widespread protests, such as the 2006 Indian anti-reservation protests, with many complaining of reverse discrimination against the forward castes (the castes that do not qualify for the reservation). Many view negative treatment of forward castes as socially divisive and equally wrong.

Caste System among Non-Hindus

In some parts of India, Christians are stratified by sect, location, and the castes of their predecessors, usually in reference to upper class Syrian Malabar Nasranis. Christians in Kerala are divided into several communities, including Syrian Christians and the so-called "Latin" or "New Rite" Christians.

Syrian Christians derive status within the caste system from the tradition that they are converted Namboodiris and Jews, who were evangelized by St. Thomas. Writers Arundhati Roy and Anand Kurian have written personal accounts of the caste system at work in their community. Syrian Christians, especially Knanaya Christians, tend to be endogamous and not to intermarry with other Christian castes.

The Latin Rite Christians were among the scheduled castes in the coastal belt of Kerala, where fishing was the primary occupation. They were actively converted by missionaries in the 16th and 19th centuries. These missionary activities were carried out by Western Latin Rite missionaries who did not understand the significance of the caste system in India; mone of the Syrian churches had participated in such activities among the scheduled castes of India because they were aware of the prejudices of the caste system. The government of India later granted this group OBC status. Very rarely are there intermarriages between Syrian Christians and Latin Rite Christians.

Anthropologists have noted that the caste hierarchy among Christians in Kerala is much more polarized than the Hindu practices in the surrounding areas, due to a lack of jatis. Also, the caste status is kept even if the sect allegiance is switched (i.e. from Syrian Catholic to Syrian Orthodox).

In the Indian state of Goa, mass conversions were carried out by Portuguese Latin missionaries from the 16th century onwards. The Hindu converts retained their caste practices. The continued maintenance of the caste system among the Christians in Goa is attributed to the nature of mass conversions of entire villages, as a result of which existing social stratification was not affected. The Portuguese colonists, even during the Goan Inquisition, did not do anything to change the caste system. Thus, the original Hindu Brahmins in Goa now became Christian *Bamons* and the Kshatriya became Christian noblemen called *Chardos*. The Christian clergy became almost exclusively Bamon. Vaishyas who converted to Christianity became *Gauddos*, and Shudras became *Sudirs*. Finally, the Dalits or "Untouchables" who converted to Christianity became *Maharas* and *Chamars*, the latter an appellation of the anti-Dalit ethnic slur *Chamaar*.

Units of social stratification, termed "castes" by many, have developed among Muslims in some parts of South Asia. Sources indicate that the castes among Muslims developed as the result of close contact with Hindu culture and Hindu converts to Islam. The Sachar Committee's report commissioned by the government of India and released in 2006 documents the continued stratification in Muslim society.

Among Muslims, those who are referred to as Ashrafs are presumed to have a superior status derived from their foreign Arab ancestry, while the Ajlafs are assumed to be converts from Hinduism, and have a lower status. In addition, the *Arzal* caste among Muslims was regarded by anti-caste activists like Ambedkar as the equivalent of untouchables. In the Bengal region of India, some Muslims stratify their society according to 'Quoms'. While many scholars have asserted that the Muslim castes are not as acute in their discrimination as those of the Hindus, some like Ambedkar argued that the social evils in Muslim society were "worse than those seen in Hindu society". The Buddhists also had a caste system. In Sri Lanka, the Rodis might have been outcast by the Sri Lankan Buddhists due to the absence of *ahimsa* (*non-violence*), a central tenet of Buddhism, among their beliefs. The writer Raghavan notes, "That a form of worship in which human offerings formed the essential ritual would have been anathema to the Buddhist way of life goes without saying; and it needs no stretch of imagination that any class of people in whom the cult prevailed or survived even in an attenuated form would have been pronounced by the sangha (i.e. the Buddhist clergy) as exiles from the social order." Savarkar believed that the status of the backward castes (e.g. Chamar) that performed non-violence only worsened. When Ywan Chwang traveled to South India after the period of the Chalukyan Empire, he noticed that the caste system had existed among the Buddhists and Jains.

Jains also had castes in places such as Bihar. For example, in the village of Bundela, there were several *"jaats"* (*groups*) amongst the Jains. A person of one *"jaat"* cannot intermingle with a Jain or another *"jaat"*. They also could not eat with the members of other *"jaats"*.

The Sikh Gurus criticized the hierarchy of the caste system. While some castes were widely perceived as being better or higher than others (e.g. Brahmins being higher than others), they preached that all sections of society were valuable and that merit and hard-work were essential aspects of life. In the Shiromani Gurdwara Prabandhak Committee, out of 140 seats, 20 are reserved for low caste Sikhs. However, the quota system has attracted much criticism due to the lack of meritocracy, since merit is considered the single most important component of winning a seat.

Baha'i Faith has grown to prominence in India, since its philosophy of the unity of humanity attracted many of the lower castes.

Bibliography

Allen, John C. and Dillman, Don: *Against All Odds: Rural Community in the Information Age*, Boulder: Westview Press, 1994.

Berry, J.K.: *Beyond Mapping, Concepts, Algorithms and Issues in GIS*, Fort Collins, CO: GIS World Books, 1993.

Beteille, Andre.: *The Backward Classes and the New Social Order*, Oxford University Press, Delhi, 1981.

Bose, N. K.: *The Scheduled Castes and Tribes and Their Present Conditions*, Calcutta, University of Calcutta, 1969.

Chaklader, Snehamoy: *Sociolinguistics: A Guide to Language Problems in India*, New Delhi, Mittal, 1990.

Clark, Michael J.: *Horizons in Physical Geography*. Basingstoke and London: Macmillan Education, 1987.

Damon, William: *Social and Personality Development, Infancy through Adolescence,* New York: Norton, 1983.

Davey, Kenneth J.: *Elements of Urban Management,* Urban Management Program Discussion, The World Bank, Washington DC, 1993.

Dennis Shirley: *Community Organizing for Urban School Reform,* Austin, University of Texas Press, 1997.

Dunkerley, Harold B.: *Urban Land Policy, Issues and Opportunities, Introduction and Overview,* Oxford University Press, New York, 1983.

Elangovan K.: *GIS, Fundamentals, Applications and Population Geography*, New India Publishing Agency, New Delhi 2006.

Gurukal, Rajan: *The Formation of Caste Society in Kerala*, Rawat Publications, New Delhi, 1994.

Haggett, Peter: *Geography: A Modern Synthesis*. New York: Harper and Row, 1975.

Jackson, Peter: *Maps of Meaning: An Introduction to Cultural Geography*. London and Boston: Unwin Hyman, 1989.

James S. Duncan: *Writing Worlds: Discourse, Text, and Metaphor in the Representation of Landscape*. London and New York: Routledge, 1992.

Kidokoro, Tetsuo: *Development Control Systems for Housing Development in Southeast Asian Cities,* Regional Development Dialogue, UNCRD, Nagoya, Japan, 1992.

Lung, Y., and van Tulder, R.: *Cars, Carriers of Regionalism,* Palgrave Macmillan, New York, 2004.

McDonnell, R.A. : *Principles of Geographical Information Systems,* Oxford University Press, Oxford, 1998.

Michael, H.: *City Form and Natural Process: Towards a New Urban Vernacular.* London: Croom Helm, 1984.

Neil, H.: *The Land of Contrasts: 1880-1901,* The American Culture 5. New York: G. Braziller, 1970.

Peter, G.: *The Slow Plague: A Geography of the AIDS Pandemic.* Oxford and Cambridge, MA: Blackwell Publishers, 1993.

Prajapati, R.V.: *A Hand Book of Geography : Realms, Regions and Concepts,* Cyber Tech Pub, Delhi, 2010.

Quaino, Massimo: *Geography and Marxism.* Translated by Alan Braley. Totowa, N.J.: Barnes and Noble, 1982.

Ray Allen: *The Far Western Frontier, 1830-1860.* The New American Nation Series. New York: Harper, 1956.

Reddy, L. Venkateswara, Narayana, M. Lakshmi : *Methods of Teaching Rural Sociology,* Discovery, Delhi, 2004.

Robert Fisher and Peter Romanofsky: *Community Organizing for Urban Social Change, A Historical Perspective,* Greenwood Press, Delhi, 1981.

Stephen, D.: *Fields of Vision: Landscape Imagery and National Identity in England and the United States.* Cambridge: Polity, 1993.

Tulder, R.: *Cars, Carriers of Regionalism,* Palgrave Macmillan, New York, 2004.

Victor, A.: *Geography of Art and Culture,* Contributions to Economic Analysis, Elsevier, 2004.

Watkins, Eric: *The Middle Eastern Environment,* London, The British Society for Middle Eastern Studie, 1995.

Worboys, Michael, and Matt Duckham.: *GIS, a Computing Perspective,* Boca Raton, CRC Press, 2004.

Index

❑❑❑